ANALYSIS OF SKILLS DEVELOPMENT FROM COMPUTER-ASSISTED TEACHING

DOUGGLAS HURTADO CARMONA

ANALYSIS OF SKILLS DEVELOPMENT FROM COMPUTER-ASSISTED TEACHING

Dougglas Hurtado Carmona

© 2011, Copyright of this edition:
Dougglas Hurtado Carmona

ISBN: 978-1-257-81756-6

Translated from:
Análisis del desarrollo de competencias desde la enseñanza asistida por computador

© 2011. ISBN: 978-1-257-81753-5

More information:

dhurtado@samartinbaq.edu.co
dougglash@yahoo.com.mx
dougglas@gamil.com

ACKNOWLEDGEMENTS

GOD Almighty.

San Martin University Foundation, especially Dr. **José Santiago Alvear,** and the Faculty of Engineering **(Jorge, Lucho, Horacio, Nelson and Karol).**

Students in the course Operating systems
Very vividly unwittingly participated in this project

AUTOR

DOUGGLAS HURTADO CARMONA

Master of Computer Systems Engineering, Computer Systems Engineering, Minor in Business Administration and Information Systems Security. IBM Certification in Information Systems Management. Diploma in Scientific Research, Development for Web Applications, Computer Security and Forensic Computing, and in education and pedagogy.

Since 2002 he works as Head of Research at the Faculty of Engineering of the San Martin University Foundation Headquarters Puerto Colombia, Barranquilla - Colombia.

National and international lecturer with 12 years of university teaching experience in the areas of Programming Objects, Data Structures Object Oriented, Systems Theory, Systems Analysis and Design, Operating Systems, Compilers, Databases, Programming Concurrent and Client / Server Java, application development for Internet, computer security, computer forensics and contingency plans.

Researcher on the topics of Information Security, Computer Forensics, General Systems Theory and dynamic systems for software engineering, and compiler theory. Developed the research "Analysis of skill development from the use of Computer Assisted Teaching" which received Special Mention Awards ACOFI 2007, and "Methodology for the development of systems based on learning objects." Creator OSOFFICE, Educational Software for Teaching Operating Systems.

He has served as Director of software development projects, systems analyst and programmer, IT project manager, information security engineer, independently has advised companies involved in building software.

TABLE OF CONTENT

1. PREPOSITION OF THE INVESTIGATIVE PROJECT..	1
INTRODUCTION	1
PROJECT DESCRIPTION..	2
Project Title...	*2*
Summary...	*2*
Motivation (Anecdote)	*2*
Interested entity...	*3*
Estimated cost and time...	*3*
RESEARCH PROBLEM...	3
Brief Description of the Problem	*3*
Problem Formulation...	*4*
JUSTIFICATION..	4
OBJECTIVES...	5
HYPOTHESIS OF THE PROJECT...	5
Type of Hypothesis...	*5*
Statement of Hypothesis...	*5*
VARIABLES..	6
Description of Variables..	*6*
Operationalizing Variables..	*7*
DESIGN METHODOLOGY...	7
Adopted Design..	*7*
Research Type..	*7*
Information Collection techniques..	*8*
Population and Sample..	*11*
Information Processing...	*11*
DELIMITATION...	12
Conceptual delimitation..	*12*
Temporal delimitation...	*12*
Spatial delimitation	*13*
2. RESULTS OF THE INVESTIGATIVE PROCESS..	15
PRESENTATION OF THE INFORMATION COLLECTED..	15
Formation of the group CATG...	*15*
Formation of the group NCATG..	*15*
Data obtained through the instrument..	*16*

Data of Academic Secretary of the Faculty...	*17*
ANALYSIS OF INFORMATION OBTAINED FROM INSTRUMENT............................	19
Description and summary data..	*19*
Calculation of Confidence Intervals...	*24*
Hypothesis Testing...	*27*
ANALYSIS OF INFORMATION PROVIDED BY THE ACADEMIC SECRETARY...........	36
Data description and summary...	*36*
Calculation of Confidence Intervals...	*37*
Comparison populations...	*40*
CONCLUSIONS AND RECOMMENDATIONS...	**45**
BIBLIOGRAPHY..	**47**

Chapter 1
PREPOSITION OF THE INVESTIGATIVE PROJECT

INTRODUCTION

In universities are transmitted -in the best case built-, knowledge without showing their practical utility. This leads to many of them are forgotten and, worse, never be taken into account in resolving problems of professional life.

By its abstract nature and difficult experimentation, some knowledge is not treated properly by the students. This creates the motivation and a sense that the concepts are on the board and are not captured in its practical approach, moreover, students do not grasp the concepts easily reference and does not stimulate their analysis. [Hurtado y Neira, 1995] In fact, these students are denied the capacity to nourish them on a professional level for higher performance in their workplace. On the other hand, inadequate skills development affects mood and self-esteem in individuals when faced with critical situations such as impact on professional work and personally.

One approach for addressing these shortcomings is: using information technology to support learning processes has been a concern that has long been investigated and proven by many people. Its inclusion within educational institutions, including the home has increased in recent years; with demand for high-quality educational software is growing. [Gómez, Galvis y Mariño, 1998]

Today we know that this use of information technology has become commoditized and evolved along with tools that take advantage of new technologies, this is speaking in the first instance, the use of ICT, **Information and Communication Technology**, in the education; from school for radio (radio school) through the use of television programs with educational support, to reach the remote education, virtual classrooms, online education, what is known as virtual education.

In a second instance the use of computers in education has specialized giving rise to the so-called learning objects according to [Aproa, 2007] "A learning object (LO) corresponds to the smallest independent structure that contains an objective, learning activity, a metadata and an evaluation mechanism, which can be developed with info-technologies (ICT) so as to facilitate reuse, interoperability, accessibility and long life". It should be noted that the tool used to perform computer-assisted instruction is composed of several learning objects.

This work is aimed at researchers, teachers and managers of educational institutions in order to motivate them to use computer-assisted instruction in any of its forms, in their courses, to show the benefit gained in the development the competence of their students to use.

This documentation represents an extract of the most important aspects of the research performed, the extract is presented in the following way: First of all, his description and brief description of the problem, then work goals, then those aspects relating to assumptions and methodology and the description of the information collected, hypothesis testing and analysis are discussed below, and finally describe the results of academic performance and sets out the conclusions and recommendations.

PROJECT DESCRIPTION

Project Title

This research work has been titled with the name of *analysis of skill development from computer-assisted instruction.*

Summary

In this project aims to calculate the proportional difference in the development of skills among students using the Computer Assisted Teaching (CAT) and those without. To this end, we propose the hypothesis that the proportional difference in the development of skills among students using the CAT and those without, to study the subject Operating Systems is 30%.

This will define the basic research project as a Quasi-Experimental design and correlational form, where they took 2 samples of 89 students, forming groups: CATG, which used computer-assisted instruction, and not used, NCALG. These groups was administered as a questionnaire and obtained partial notes on the subject. To obtain the results, we evaluate the hypothesis and compared the groups formed in the development of skills and academic performance.

Keywords: Skills Development, Computer Aided Teaching, Engineering, Operating Systems, Software education.

Motivation (Anecdote)

In mid-1999 was conceived within the Faculty of Engineering, a controversy between two teachers with the convenience of using the Computer Assisted Teaching for better development of student performance to take the course Operating Systems. Because of this academic controversy and to demonstrate the great advantages of computer in teaching, it was decided to conduct this research project.

Interested entity

The entity concerned is the Faculty of Engineering of the San Martin University Foundation Puerto Colombia headquarters in the city of Barranquilla, Republic of Colombia.

Estimated cost and time

An estimated total cost of seventy-six millions three hundred eighty-six thousand seven hundred and fifty pesos Colombian with 00 cents ($ 76'386.750). Development time of this Research corresponds eight (8) semesters academic from second 1999 until first 2003, which are 128 weeks approximately from project approval.

RESEARCH PROBLEM

Brief Description of the Problem

In the first curriculums -ran the 1997 to 2003- the program of Computer science Engineering used in less than 10% of their teaching subjects from the use of educational software and / or application (CAT). This is possibly a result of the following situations:

Lacking or inappropriate Integration of CAT in the culture of teaching in the institution: Many professors, deans and managers have very weak notions of teaching contribution representing the CAT. "Know" that works because they get the news from abroad, but not displayed as part of their culture of teaching in higher education. Just be content with the "idea" that will someday be widely used: "This is the Tool of the Future" they say. [Hurtado y Neira, 1995]

Little awareness of the benefits in using a tool CAT in higher education: By not integrating CAT in the culture of teaching, it is clear that is not used, and therefore there will be no training plans and training, further promoting the development of educational software projects will be null and void is the acquisition of such tools. With all the above is even larger gap of ignorance of the educational benefits of the CAT.

Wrong conception of what an educational software in conjunction with the association "expenditure" without monetary gain: The majority of managers perceive the software generally, including education, as an "object or an abstract entity" very accessible to their understanding represents one more expense. On the one hand, this conception generates a rejection of the unknown, and secondly, a "cost" should be avoided as much as possible. Consequently, managers do not invest in such tools as "unknown" have educational potential. [Hurtado y Neira, 1995]

Poor infrastructure in data centers: Thus, as the software regarded as an "object or an abstract entity" very accessible to his understanding that represents an expense over the hardware is regarded in the same way.

Management to computer fraud fears: the paradigm "I do not know much about it and I can cheat" over the institutions fallible computer deceit and unscrupulous people say what they want to hear.

Poor computer literacy of the directives: higher education institutions do not have positions in your organization "vice" for the management of IT projects, it is then, managers with no experience or computer literacy are responsible for these tasks. No training

Lack of contextualization of the subjects aimed at the utilization of the CAT: Many subjects in the Systems Engineering Program remain "on board" and are not captured in its practical approach, moreover, students may not easily grasp the concepts referencing and analysis is not encouraged.

In fact, students of computer science Engineering (and other professions) are deprived of a tool that would foster a professional level for higher performance in their workplace. Concepts that could be assimilated in better shape using educational software, then not used, poorly focused and poorly treated.

Finally, in addition to not acquiring the tools CAT, does not generate motivation for the design and development of educational software projects in the students.

Problem Formulation

This project seeks to answer the following question:

What is the difference in the level of learning among students of the Engineering of computer science program to study the subject operational systems through the CAT and without CAT?

JUSTIFICATION

This research aims to exclusively find the difference in the level of student learning Computer science Engineering Program, Faculty of Engineering at San Martin University Foundation using CAT. and not, as well as related aspects.

This will enable managers, deans and even teachers, encourage the creation of an information culture within the institution, and create procedures for the evaluation and acquisition of educational software, also generating initiatives for the design and construction projects CAT tools. in college. Similarly, students can conceptualize the concepts and practical use in their professional development.

With the socialization of the results is intended that institutions are aware of the benefits of the Computer-Assisted Teaching, and so create or enhance the integration of this educational culture.

Another objective is to change the idea that the educational software product is an expense; with concrete actions to improve the quality of education have an impact on institutional prestige to turn back to other aspects as monetary gain enrollment, access resources for teaching and research, etc:

OBJECTIVES

The general objective which seeks to achieve in this investigation is stated as follows:

Calculate the proportional difference in the development of skills among students using the CAT and not use it, to study the subject Operating Systems, to develop strategies to, in part, to teachers using the CAT in pedagogy class, and in part to encourage the generation of construction projects Educational Software.

In order to tackle with the overall goal described above, must meet the following goals.

- *Define Operating Systems topics that serve as the basis for conducting the research.*
- *Select educational software and / or application of knowledge applicable to the area defined Operating System to be used in the process of establishing the differences in levels of student learning.*
- *Design of data collection instruments.*
- *Select the experimental sample.*
- *Apply information gathering tools to the selected sample.*
- *Test the hypothesis of the project and analyze the results to make, pictorial*

HYPOTHESIS OF THE PROJECT

Type of Hypothesis

Given that the current project is framed to compare the behavior of Students using the CAT and those without, to attend Operating Systems Course, we certainly say that the formulation type Hypothesis of this project is to **group differences.**

Statement of Hypothesis

Under the objective sought with this research is to know if you can accept the following hypothesis:

H1: *The proportional difference in the development of skills among students using the CAT and not use it, to study the subject Operating Systems, is 30%.*

VARIABLES

Description of Variables

To verify the hypothesis proposed in the draft the following variables: *Using Computer-assisted Teaching*; and *Skills Development*, which are described below:

Using Computer-Assisted Teaching

The Use of Computer-Assisted Teaching is, as its name indicates, the use or not of a computational tool to support teaching and learning process in the computer science engineering program in the subject Operational Systems for selected the experiment.

Behavior "causal" or "influences" that characterizes the variable Using Computer-Assisted Teaching defines its character as **Independent**. Its dimension is teaching the course Operating Systems. Has a single indicator known **use**, takes discrete values and Boolean (True or False).

Skills Development

This feature describes the state of performance of the knowledge, skills and values result of the learning process towards effective development of a professional activity related to the operational systems.

The hypothesis seeks to understand the relationship between use of Computer Assisted Teaching and the effect it has to develop skills, which is why this is classified as a **dependent** variable in the first (Use of Computer Assisted Teaching).

The variable Skills Development has three (3) dimensions: The Interpretative, the Argumentative and the Proposals. The **Interpretative** framed achievements based on the ability to make sense from either a text, a proposition, a problem, etc. The **Argumentative**, based on the extent of achievement orientation to account for a statement, articulate concepts and theories to support, justify, build relationships, demonstrate and conclude. Finally, the **Proposals**, based on achievements such as: proposing hypotheses, solve problems and build alternative solutions.

In the three dimensions of this variable have an indicator called the **hit ratio**. This indicator shows actual values between 0 and 1 that are the result of the ratio of correct hits and the number of tests. The hit ratio determines a qualitative assessment as follows:

- **Poor**: When we get less than 60% of the hits. [0% -59%]
- **Acceptable**: When we get between 60% to 79% of the hits. [60% -79%]
- **Good**: When we get between 80% to 90% of the hits. [80% -90%]
- **Excellent**: When you get hit over 90%. [91% -100%]

Operationalizing Variables

The process of operationalization, i.e. the empirical consequences of the variables described in the following table (Table 1) taking into account its dimensions and performance indicators

TABLE 1. OPERATIONALIZING VARIABLES

Variables	Dimension	Indicators
Using Computer-Assisted Teaching	Teaching the course Operating Systems	USE
Skills Development	1. Interpretative	Hit ratio
	2. Argumentative	Hit ratio
	3. Proposals	Hit ratio

DESIGN METHODOLOGY

Adopted Design

Research design is **Quasi - Experimental**, because it deliberately manipulates the independent variable use of CAT in order to observe the behavior of the dependent variable skills Development, also because the comparison groups are not randomly selected or matched, but these groups are already formed before applying the experiment, i.e. they are intact groups.

We may add that the basis of the experiment is to apply the instrument to a single subject courses, where CAT use and others what not uses in different semesters

Research Type

The type of research is **basic** and that this project is undertaken the task of obtaining knowledge or principles in order to create a point of support for troubleshooting. In addition, because this project has an immediate goal theory.

Moreover, based on the type of experiment, we can say that this project presents the form of **correlational** research are intended to show the relationship between variables.

Information Collection techniques

Primary data collection techniques

The source of primary gathering that will be used project presently is the **Survey**, with **Experimental** modality, using the Instrument **Questionnaire**.

Description of the instrument

The instrument (questionnaire) was divided into five (5) sub themes: Fundamentals of Operating Systems, Process Management, Memory Management, File Management and secondary storage, and communication processes and process control. Which in turn are classified according to the type of question the kind of competition to be evaluated. The instrument is as follows:

TABLE 2. RESEARCH TOOL

Sub theme I: **Fundamentals of Operating Systems**
A. Interpretative competences
1. Define the concept of Operational System 2. Name the Classification of operation system 3. Describe the structure of Operating Systems
B. Argumentative competences
4. Arguing the History of Operating Systems 5. Make a comparative table of the types of Operating Systems. Give examples of current market 6. Find the differences between monolithic and hierarchical structure of Operating Systems.
Sub theme II: **Process Management - Planning Processor**
A. Interpretative competences
7. Define the concept of Process 8. Define the PCBs 9. Define to the system queues 10. Define the evaluation indexes
B. Argumentative competences
11. Describe the importance of the concept of multitasking 12. Describe the importance of planning processes 13. Make a graphic description of the states of processes 14. Build a comparison chart describing the functioning of the policies FCFS, SJF, Round Robin, by priority, appropriate SJF, appropriate priorities

C. Propositional competences
15. Building a method or function that simulates the operation of the FCFS policy
16. Building a method or function that simulates the operation of the policy by appropriate priorities
17. Find the best planning, noting the changes in the tails of the system, comparing the average waiting time and creating the corresponding Gantt charts, using the following data:

Arrival Time	Process	Cpu cycles
0	P1	2
0	P2	5
1	P3	4
2	P4	8
5	P5	5
7	P6	4

Sub theme III:
Memory Management

A. Interpretative competences
18. Define the concept of addressing
19. What is the Mono programming?
20. What is Multi programming?
21. What are the fundamental concepts of contiguous memory management?
22. Define the paging concept
23. Define segmentation

B. Argumentative competences
24. Point out the differences between paging and segmentation
25. Why is it important to protect your memory?
26. Make a table comparing the policies of partitions of fixed size and variable.
27. What is the difference between internal and external fragmentation?
28. Describe the importance of planning the main memory
29. Why is important the virtual memory?
30. Build a comparison chart describing the functioning of page replacement policies, FIFO, LRU, Optimal and Clock
31. What is the importance of Page Faults?

C. Propositional competences
32. Building a method or function that simulates the operation of the page replacement policy FIFO
33. Find the best page replacement planning for the following applications: 2, 3, 4, 2, 5, 6, 5, 3, 6, 7, 8, 9, 2, 5, 7, 6 3, 7. using three frames of the system. Write down the changes in the frames and page faults.

Sub theme IV:
File Management and secondary storage

A. Interpretative competences

34. What is the structure of information?
35. What are the methods of access to information?
36. What is the allocation of free space?
37. What controls the space?

B. Argumentative competences
38. Point out differences between the direct, indexed and sequential information access
39. Make a chart describing hardware information
40. What is the difference between directory and file device?
41. Describe the importance of the directory tree
42. Make a comparative table of the planning algorithms disk access (FCFS, SSTF, SCAN, C-Scan)

C. Propositional competences
43. Building a method or function that simulates the operation of the disk access policy SSTF
44. Find the best disk access planning for the following applications: 28, 32, 4, 23, 51, 68, 55, 33, 63, 76, 83, 90, 27, 55, 74, 46 34, 73, used with an initial position 45.
45. Taking data from the previous year to create a new planning method of disk access with the lowest access time.

Sub theme V:

Communication processes and process control

A. Interpretative competences
46. What is the concurrency?
47. What is a concurrent program?
48. What is a shared variable?
49. Define the concept of semaphore
50. What class is it used to carry out concurrent programs in java?
51. How do they communicate the processes?
52. How processes are they controlled?
53. What is the synchronization of processes?

B. Argumentative competences
54. Identify problems of concurrency
55. Describe the functional differences between shared variable and semaphore
56. Discuss the principles and benefits of concurrency
57. Describe the advantages and disadvantages to synchronize processes
58. Describe the structure of the execution of a thread or thread

C. Propositional competences
59. Build a model - general staff of the classes in Java for concurrent programs
60. Build a model - the general staff in Java classes shared variables
61. Design and build a Java program a computer sales store in which there are "n" providers while stocks added to the inventory between 1 and 5 teams and m buyers decrement stocks between 1 and 2 teams, in addition eventually there is a thief steals a computer.

Population and Sample

The population is made up of students enrolled in Computer Science Engineering program, Faculty of Engineering of the San Martin University Foundation. To calculate the size of the sample using the formula for finite populations or known [Berenson, 1996; page 350]:

$$n = \frac{Z^2 * p * q * N}{(N-1) * e^2 + Z^2 * p * q}$$

Where: n: Sample size; Z^2: confidence -Level of trust, p: positive variability; q: negative Variability, N: population size, (N-1): Level of accuracy, e^2: MPE (margen permited error)

Performing the calculation with a population size of 230 students (Number of students in college in the 1999-2 academic semester) in the program in half for each stock is 115, a confidence level of 95% (Z = 1.96) and a margin of error of 5%, and percentages of positive and negative variation of 50% we have:

n = [$(1.96)^2$*(0.5)(0.5)*115] / [(114)*(0.05)2 + $(1.96)^2$*(0.5)(0.5)]
n = 110.446/ [0.285+ 0.9604]
n = 110.446/ 1.2454
n = 88.6831

We conclude that we need **89** students to be representative of each population.

Information Processing

For information processing is taken into account the following:

1. Students enrolled in each semester to study the subject Operating Systems will be taken as part of the sample.
2. The instrument will be applied to every student in the sample.
3. Semesters were selected in which to apply the CAT and those without. Can be consecutive or not.
4. After obtaining the data are classified and tabulated into two groups according to use or not of the CAT.
5. Statistical procedure is used to test hypotheses.
6. The results will be displayed in graphical form.

DELIMITATION

Conceptual delimitation

The themes addressed in the experiment refer to the topics of traditional operational systems to the following special items described in Table 3:

TABLE 3. CONCEPTUAL DELIMITATION

OPERATING SYSTEM BASICS [Milenkovic, 1997] [Silberschatz, 2006] [Tanenbaum, 2003]
• EVOLUTION OF OPERATING SYSTEMS • STRUCTURE OF OPERATING SYSTEMS. Monolithic structure, hierarchical structure, Virtual Machine, Client-Server.
MANAGEMENT PROCESS [Silberschatz, 2006] [Milenkovic, 1997] [Tanenbaum, 2003]
• BASICS. Process Concept. Types of processes. Exceptions. • THE PROCESS CONTROL BLOCK (PCB). Process status. Active states. Inactive states. State transitions. Operations processes. Priorities. • PLANNING PROCESS. Planning concept. Objectives. Criteria. Measures. Planning algorithms. First come, first served (FCFS). Round-Robin (RR). The next process, the shortest (SJF). Next process, the shortest remaining time (SRT). Priority. Near the highest response rate (HRN).
MEMORY MANAGEMENT [Stallings, 2005] [Tanenbaum, 2003]
• BASICS. Introduction. Addressing. Address assignment. Storage hierarchy. Monoprogramming. The dedicated memory. Memory Division. The resident monitor. Memory protection. Address reassignment. Exchange storage. Multiprogramming. Memory protection. Contiguous fixed-size partitions. Contiguous partitions of varying size. • PLANNING MEMORY. Planning concept. Planning Policy. Paging. Memory management. Performance. Cache. Associative registers. Shared pages. Segmentation. Hardware segmentation. Rendimiento.Sistemas combined. • VIRTUAL MEMORY. Page load request. Replacement pages. Replacement algorithms. FIFO replacement algorithm. LRU algorithm. Other algorithms. Criteria for replacement of pages. Memory allocation. Location of the processes. Page fault frequency.
FILE MANAGEMENT AND STORAGE SECONDARY [Silberschatz, 2006] [Milenkovic, 1997]
• BASICS. Introduction. Information structure. Physical support for the information. Physical and logical records. Device directory. File directories. Level directories. Two-level directories. Multilevel structures. Directory trees. Other directory structures. • PLANNING DISK ACCESS. Access methods. Sequential access. Shortcut. Shortcut indexed. Space allocation. Control of space. Storage space allocation. Contiguous allocation. Linked allocation. Indexed allocation. Planning algorithms. First come, first access (FCFS) First, the lowest seek time (PBS). (SCAN) Scanning circular (C-SCAN)

Temporal delimitation

The development time of this research is for eight (8) academic semesters at the Faculty of Engineering from the second from 1999 until the first of 2003, which are approximately 128 weeks after the approval of the project.

Spatial delimitation

This research was conducted in the Faculty of Engineering at San Martin University Foundation Headquarters Caribbean, Km 8 route to Puerto Colombia.

14 Analysis of skills development from computer-assisted teaching

Chapter 2
RESULTS OF THE INVESTIGATIVE PROCESS

PRESENTATION OF THE INFORMATION COLLECTED

To perform the experiment were required to submit two independent populations, first, take the subject property of operational systems in the FUSM and one of them used the CAT and the other not, and second, to have software support for education well as appropriate facilities.

The software selected was OSOffice 3.0, which applies to the teaching of operational systems. The groups were called CALG and NCATG the first receive computer assisted instruction, and the second not.

Formation of the group CATG

CATG group was formed with students from different academic semesters as in Table 4 is described. In the 2003-1 period were randomly selected to complete the required number of 89 sample items.

TABLE 4.   FORMATION OF THE GROUP CATG

Year	Semester	No. Students
1999	Second	9
2000	First	17[1]
2001	First	21
2003	First	42

Formation of the group NCATG

Similarly NCATG group complies with students of different semesters. Your description can be seen in Table 5. In the 2002-2 period were randomly selected to complete the required number of 89 sample items.

---

[1] Parte de este salón se ubicó en ambos grupos

TABLE 5. FORMATION OF THE GROUP NCATG

Year	Semester	No. Students
2000	First	8
2000	Second	17
2001	Second	8
2002	First	23
2002	Second	33

Data obtained through the instrument

The instrument was administered to a total of 178 students, half constituent CATG group and half to NCATG. The student responses CATG group described in Table 6 and NCATG Group in Table 7. In these tables the column means the sequential Stud student in the group; Hits is the number of correct answers in the instrument of 61 questions, and Prop. is the proportion of number of hits on the total number of questions.

TABLE 6. DATA OBTAINED THROUGH THE INSTRUMENT - CATG

Stud	Hits	Prop.	Stud	Hits	Prop	Stud	Hits	Prop.	Stud	Hits	Prop.
1	56	0.9180	23	57	0.9344	45	57	0.9344	67	60	0.9836
2	58	0.9508	24	59	0.9672	46	57	0.9344	68	58	0.9508
3	59	0.9672	25	56	0.9180	47	59	0.9672	69	59	0.9672
4	58	0.9508	26	57	0.9344	48	59	0.9672	70	58	0.9508
5	57	0.9344	27	58	0.9508	49	55	0.9016	71	55	0.9016
6	56	0.9180	28	59	0.9672	50	57	0.9344	72	53	0.8689
7	58	0.9508	29	58	0.9508	51	59	0.9672	73	57	0.9344
8	57	0.9344	30	59	0.9672	52	58	0.9508	74	56	0.9180
9	57	0.9344	31	55	0.9016	53	54	0.8852	75	55	0.9016
10	57	0.9344	32	59	0.9672	54	58	0.9508	76	58	0.9508
11	54	0.8852	33	56	0.9180	55	59	0.9672	77	60	0.9836
12	56	0.9180	34	58	0.9508	56	54	0.8852	78	54	0.8852
13	58	0.9508	35	59	0.9672	57	58	0.9508	79	58	0.9508
14	60	0.9836	36	58	0.9508	58	57	0.9344	80	59	0.9672
15	59	0.9672	37	60	0.9836	59	59	0.9672	81	60	0.9836
16	58	0.9508	38	58	0.9508	60	55	0.9016	82	53	0.8689
17	56	0.9180	39	59	0.9672	61	59	0.9672	83	60	0.9836
18	57	0.9344	40	56	0.9180	62	58	0.9508	84	57	0.9344
19	56	0.9180	41	58	0.9508	63	56	0.9180	85	58	0.9508
20	59	0.9672	42	59	0.9672	64	58	0.9508	86	58	0.9508
21	58	0.9508	43	59	0.9672	65	57	0.9344	87	57	0.9344
22	59	0.9672	44	58	0.9508	66	57	0.9344	88	59	0.9672
									89	59	0.9672

TABLE 7. DATA OBTAINED THROUGH THE INSTRUMENT - NCATG

Stud	Hits	Prop.	Stud	Hits	Prop	Stud	Hits	Prop.	Stud	Hits	Prop.
1	35	0.5738	23	42	0.6885	45	41	0.6721	67	42	0.6885
2	41	0.6721	24	37	0.6066	46	41	0.6721	68	44	0.7213
3	37	0.6066	25	38	0.6230	47	47	0.7705	69	38	0.6230
4	44	0.7213	26	32	0.5246	48	46	0.7541	70	43	0.7049
5	45	0.7377	27	43	0.7049	49	41	0.6721	71	41	0.6721
6	40	0.6557	28	43	0.7049	50	34	0.5574	72	41	0.6721
7	36	0.5902	29	43	0.7049	51	41	0.6721	73	38	0.6230
8	41	0.6721	30	37	0.6066	52	40	0.6557	74	40	0.6557
9	46	0.7541	31	40	0.6557	53	37	0.6066	75	36	0.5902
10	35	0.5738	32	38	0.6230	54	43	0.7049	76	43	0.7049
11	37	0.6066	33	45	0.7377	55	51	0.8361	77	35	0.5738
12	42	0.6885	34	37	0.6066	56	43	0.7049	78	35	0.5738
13	38	0.6230	35	45	0.7377	57	43	0.7049	79	38	0.6230
14	34	0.5574	36	48	0.7869	58	43	0.7049	80	34	0.5574
15	44	0.7213	37	38	0.6230	59	36	0.5902	81	41	0.6721
16	42	0.6885	38	40	0.6557	60	34	0.5574	82	38	0.6230
17	33	0.5410	39	39	0.6393	61	32	0.5246	83	44	0.7213
18	38	0.6230	40	44	0.7213	62	38	0.6230	84	46	0.7541
19	43	0.7049	41	36	0.5902	63	39	0.6393	85	37	0.6066
20	36	0.5902	42	46	0.7541	64	36	0.5902	86	37	0.6066
21	43	0.7049	43	37	0.6066	65	48	0.7869	87	38	0.6230
22	35	0.5738	44	39	0.6393	66	37	0.6066	88	39	0.6393
									89	44	0.7213

Data of Academic Secretary of the Faculty

The final grades achieved by students in both groups CATG and NCATG in the course of Operating Systems shown in Tables 8 and 9 respectively. The description of the meaning of each column of these tables is:

Student: The student in the group sequential
Period: Academic Year in which the note was obtained
Final: Final note (from 0.0 to 5.0) earned by the student in the subject

TABLE 8. FINAL NOTES ON RAW DATA - CATG

Stud	Period	Final	Stud	Period	Final	Stud	Period	Final	Stud	Period	Final
1	011	2.80	23	031	4.42	45	031	3.05	67	992	3.74
2	011	4.07	24	031	4.14	46	031	3.17	68	992	3.39
3	011	3.70	25	031	3.39	47	031	3.22	69	992	3.57
4	011	3.42	26	031	4.79	48	031	3.59	70	992	4.49
5	011	4.07	27	031	3.68	49	031	4.24	71	992	3.42
6	011	3.30	28	031	3.29	50	031	3.05	72	992	3.61
7	011	3.93	29	031	3.05	51	031	4.09	73	001	3.48
8	011	3.02	30	031	4.10	52	031	3.25	74	001	3.48
9	011	4.40	31	031	3.46	53	031	3.39	75	001	3.54
10	011	3.30	32	031	3.00	54	031	3.05	76	001	3.54
11	011	3.55	33	031	3.22	55	031	3.06	77	001	3.63
12	011	3.08	34	031	4.94	56	031	3.40	78	001	3.63
13	011	3.38	35	031	3.32	57	031	3.60	79	001	3.66
14	011	4.85	36	031	3.02	58	031	3.20	80	001	3.72
15	011	3.60	37	031	3.22	59	031	3.50	81	001	3.78
16	011	3.90	38	031	3.47	60	031	2.50	82	001	3.78
17	011	4.70	39	031	3.36	61	031	3.10	83	001	3.84
18	011	4.52	40	031	4.27	62	031	2.00	84	001	3.90
19	011	4.70	41	031	3.65	63	031	4.20	85	001	4.00
20	011	3.32	42	031	3.15	64	992	3.50	86	001	4.00
21	011	3.58	43	031	3.64	65	992	3.88	87	001	4.00
22	031	4.30	44	031	4.12	66	992	4.64	88	001	4.23
									89	001	4.46

TABLE 9. FINAL NOTES ON RAW DATA - NCATG

Stud	Period	Final	Stud	Period	Final	Stud	Period	Final	Stud	Period	Final
1	002	3.58	23	012	3.14	45	021	3.38	67	022	3.22
2	002	3.60	24	012	3.53	46	021	3.03	68	022	4.15
3	002	3.25	25	012	2.83	47	021	3.08	69	022	3.04
4	002	3.24	26	021	3.30	48	021	3.73	70	022	2.97
5	002	3.00	27	021	3.02	49	022	2.59	71	022	3.10
6	002	3.20	28	021	2.99	50	022	2.70	72	022	3.10
7	002	4.49	29	021	2.99	51	022	4.24	73	022	2.70
8	002	3.43	30	021	1.70	52	022	3.55	74	022	4.00
9	002	3.00	31	021	3.04	53	022	3.05	75	022	2.70
10	002	3.05	32	021	3.46	54	022	3.10	76	022	3.02
11	002	3.78	33	021	2.98	55	022	3.68	77	022	2.22
12	002	3.58	34	021	3.04	56	022	2.46	78	022	3.13
13	002	3.66	35	021	3.32	57	022	3.37	79	022	4.15
14	002	3.18	36	021	3.15	58	022	0.54	80	022	3.04
15	002	3.83	37	021	2.99	59	022	0.00	81	022	4.00
16	002	4.69	38	021	3.32	60	022	3.31	82	001	3.42
17	002	3.69	39	021	3.02	61	022	3.37	83	001	2.88
18	012	3.02	40	021	3.07	62	022	3.00	84	001	2.88
19	012	3.04	41	021	3.61	63	022	3.10	85	001	2.88
20	012	2.95	42	021	3.06	64	022	3.01	86	001	2.98
21	012	3.17	43	021	3.02	65	022	2.22	87	001	3.03
22	012	3.20	44	021	3.98	66	022	3.13	88	001	4.48
									89	001	4.52

ANALYSIS OF INFORMATION OBTAINED FROM INSTRUMENT

Description and summary data

The data obtained through the instrument of each group was calculated the proportion, its variance and standard deviation, which are summarized in Table 10. [Berenson, 1996]

TABLE 10. INSTRUMENT DATA SUMMARY

Group	Total Problems	Total Hits	Mean Proportion	Variance Proportion	Deviation Proportion
CATG	5429	5119	**0.94289924**	0.00073363	0.02708555
NCATG	5429	3555	**0.65481672**	0.00432655	0.06577649

Other measures of central tendency and dispersion are listed in Table 11:

TABLE 11. MEASURES OF CENTRAL AND DISPERSION

Group	Median	Mode	Half range	Range
CATG	0.9508	0.9508	0.9226	0.8688 – 0.9836
NCATG	0.6557	0.6230	0.6803	0.5246 – 0.8360

While the frequency distribution CATG group presented in Table 12 and Figure 1, the frequency distribution of NCATG Group is presented in Table 13 and Figure 2.

TABLE 12. FREQUENCY DISTRIBUTION OF GROUP CATG

Value proportion	Frequency
0.8689	2
0.8852	4
0.9016	5
0.9180	10
0.9344	16
0.9508	24
0.9672	22
0.9836	6

20 Analysis of skills development from computer-assisted teaching

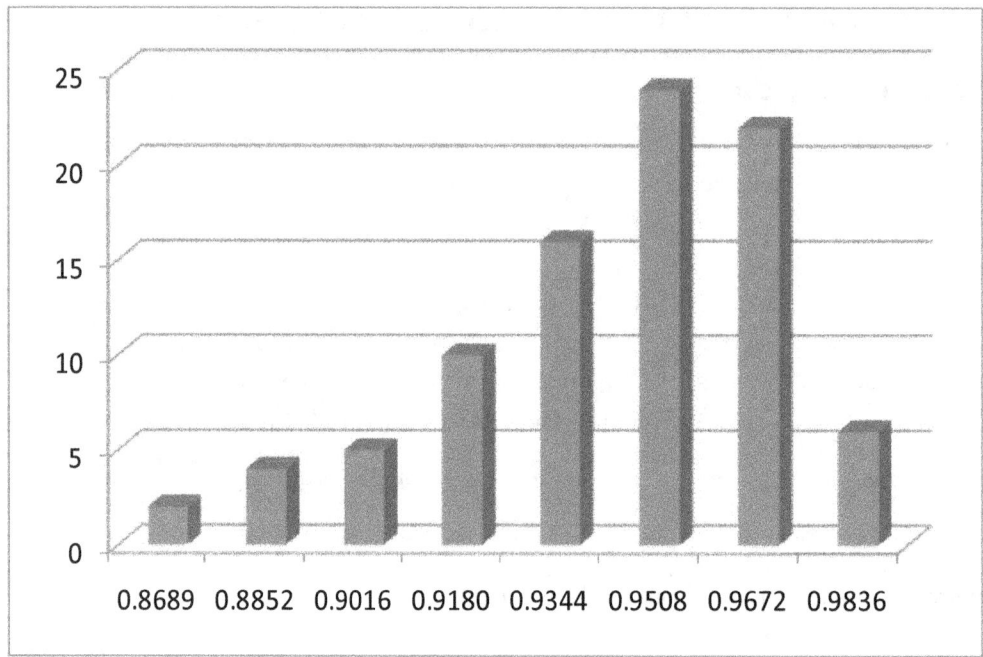

FIGURE 1. CATG Group Frequency Distribution

Likewise, questions on the instrument classified as Performing capacity interpretatives (25 questions), argumentative (25 questions) and purpose (11 questions) were classified by calculating the ratio group. (See Tables 14 and 15)

TABLE 13. FREQUENCY DISTRIBUTION OF GROUP NCATG

Value proportion	Frequency	Value proportion	Frequency	Value proportion	Frequency
0.5246	2	0.6230	11	0.7213	6
0.5410	1	0.6393	4	0.7377	3
0.5574	4	0.6557	5	0.7541	4
0.5738	5	0.6721	9	0.7705	1
0.5902	6	0.6885	4	0.7869	2
0.6066	10	0.7049	11	0.8361	1

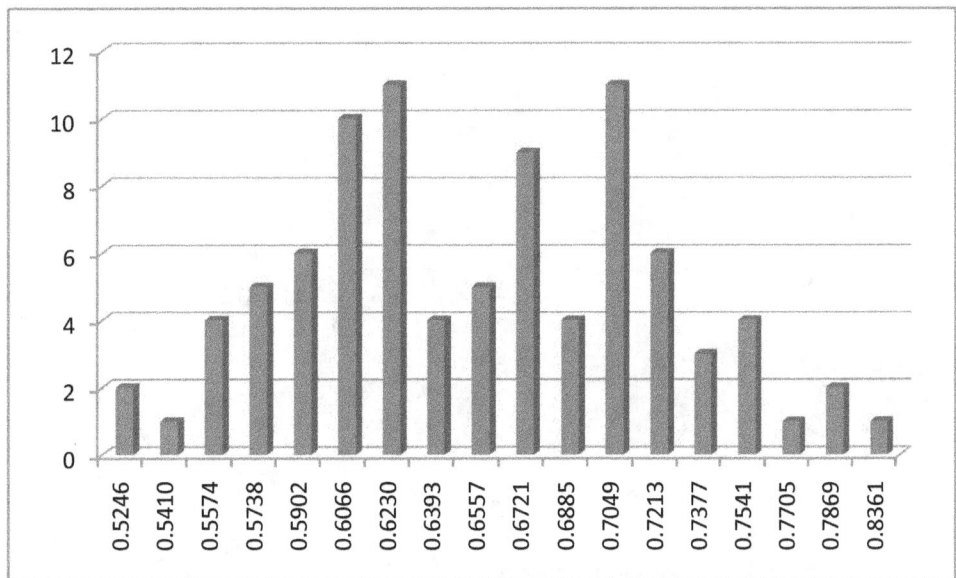

FIGURE 2. NCATG Group Frequency Distribution

TABLE 14. CLASSIFIED DATA SUMMARY FOR SKILLS - CATG GROUP

Skill	Total Questions	Total Students	Total Problems	Hits	Proportion
Interpretative	25	89	2225	2098	0.9429
Argumentative	25	89	2225	2087	0.9380
Proposals	11	89	979	934	0.9540

TABLE 15. CLASSIFIED DATA SUMMARY FOR SKILLS - NCATG GROUP

Skill	Total Questions	Total Students	Total Problems	Hits	Proportion
Interpretative	25	89	2225	1458	0.6553
Argumentative	25	89	2225	1444	0.6490
Proposals	11	89	979	653	0.6670

Figure 3 shows comparatively the proportions of the groups according to the type of skills.

22 Analysis of skills development from computer-assisted teaching

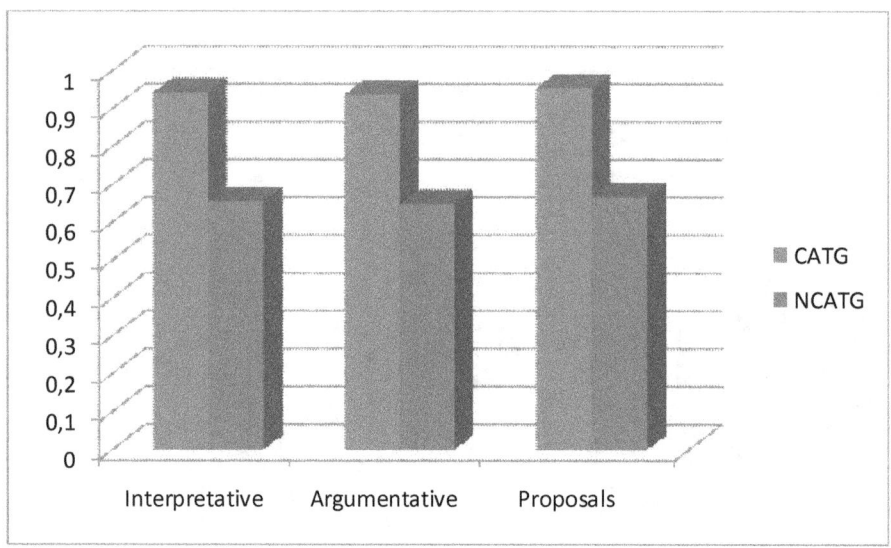

FIGURE 3. Comparison by type of skill

In Tables 16 and 17 abstracts describing the behavior of the two groups, according to the sub theme of Operational Systems.

TABLE 16. SUMMARY DATA FOR SUB THEMES - GROUP CATG

Sub Theme	Total Questions	Total Students	Total Problems	Hits	Proportion
Fundamentals of Operating Systems	6	89	534	498	0.9326
Process Management	11	89	979	920	0.9397
Memory Management	16	89	1424	1342	0.9424
File Management and secondary storage	12	89	1068	1009	0.9448
Comm. processes and process control	16	89	1424	1350	0.9480

TABLE 17. SUMMARY DATA FOR SUB THEMES - GROUP NCATG

Sub Theme	Total Questions	Total Students	Total Problems	Hits	Proportion
Fundamentals of Operating Systems	6	89	534	348	0.6517
Process Management	11	89	979	650	0.6639
Memory Management	16	89	1424	923	0.6482
File Management and secondary storage	12	89	1068	701	0.6564
Comm. processes and process control	16	89	1424	933	0.6552

Figure 4 shows the comparison of proportions of the two groups according to the sub theme of the question relating to operational systems.

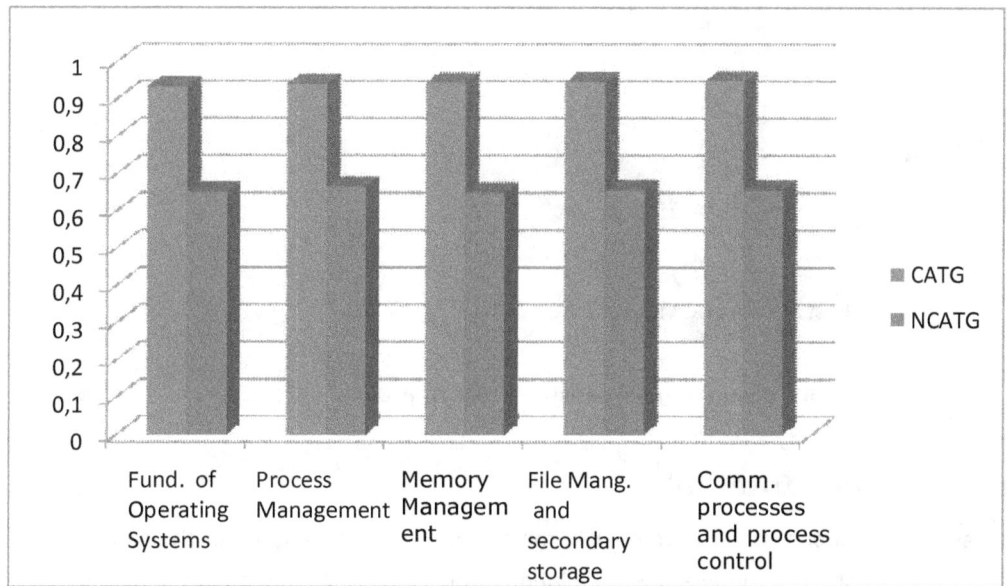

FIGURE 4. Comparison by type of skill

As in Figure 5 and 6 shows the comparative performance of the top five responses by each group.

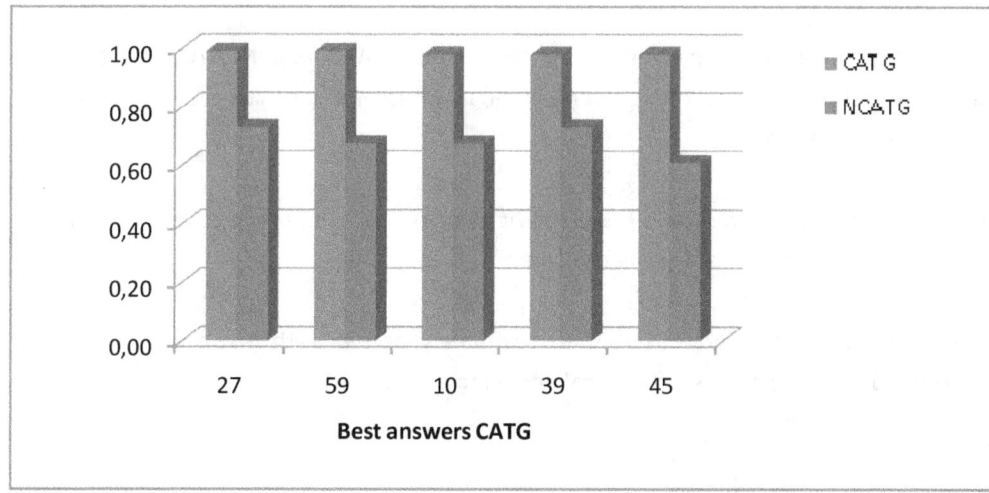

FIGURE 5. Comparing best answers CATG group

24 Analysis of skills development from computer-assisted teaching

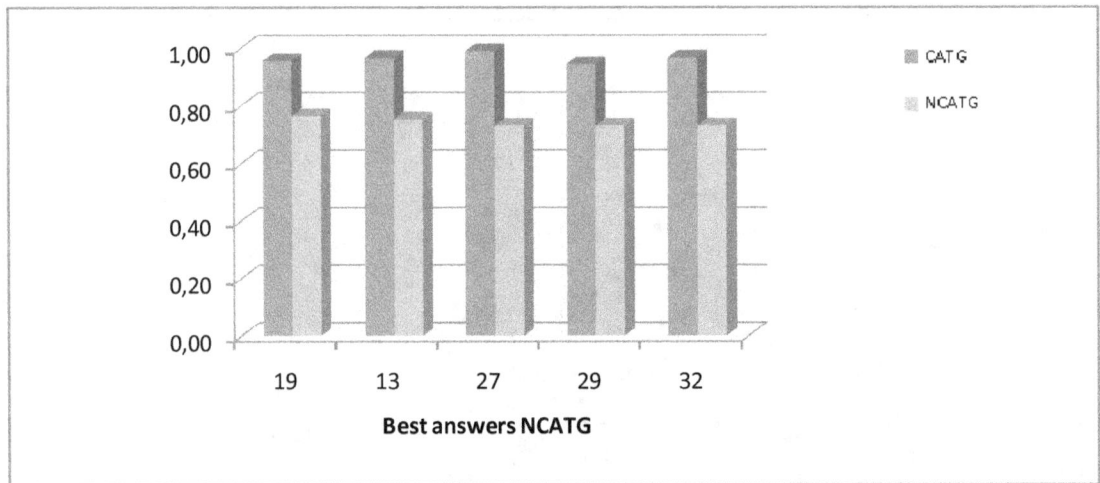

FIGURE 6. Comparing best answers NCATG group

Calculation of Confidence Intervals

In the statistical process of populations is crucial to bear in mind that the point estimate of the sample mean only gives us an approximation of the average population, which varies from sample to sample, so it is necessary to have a more accurate estimate actual characteristics of the population. This is why you should develop an interval estimate, taking the sampling distribution of the mean, real mean of the population.

Statistic to be used

To calculate the confidence intervals of the proportions of the groups CATG and NCATG, use the formula for confidence interval estimation for proportions [Berenson, 1996, Section 10.4. Formula 10.3. page 341], where we consider the following situation:

Let X be a binomial variable with parameters n and p. A binomial variable is the number of successes in n trials, in each trial the probability of success (p) is the same.

If n is large and p is not close to 0 or 1 (np> 5) X is approximately normal with mean n * p and variance n * p * q (where q = 1 - p) and can use statistical sampling proportion)

$$\hat{p} = \frac{X}{n}$$

Which is also approximately normal, with standard error given by:

$$\sqrt{\frac{pq}{n}}$$

Consequently, a confidence interval for p to 100 (1 - α)% is:

$$\hat{p} \pm z_{\alpha/2}\sqrt{\frac{pq}{n}}$$

Where:
n = the sample size
p = the proportion of correct
q = 1 - p
Z = confidence level

Which stands as the performance constraint on the premise that the number of problems for the ratio should be at least 5, which is reflected in the following mathematical expression:

$$n * p > 5 \text{ y } n * (1-p) > 5$$

Let's see if the samples of the groups meet the rule:

P_{CATG} = X/n = 0.9428

P_{NCATG} = X/n = 0.6548

For the group CATG

$P_{CATG} * n$ = 0.9428 * 5429 = 5118.4612

$(1-P_{CATG}) * n$ = 0.0572 * 5429 = 310.5388

As you can see both results are greater than 5.

For the group NCATG

$P_{NCATG} * n$ = 0.6548 * 5429 = 3554.9

$(1-P_{NCATG}) * n$ = 0.3452 * 5429 = 1874.09

Likewise, the results are greater than 5.

We can conclude that you can use the statistic proposed for both samples. To make the graphs will be used in Java program **Descartes**, created by José Luis Abreu in the project that bears the same name as the Ministry of Culture and Education of the Spanish government [MinEdu y Ciencia-GobEspañol, 2007]

Confidence interval calculation CATG group

To calculate the confidence interval CATG group have the following information:

n = 5429

X = 5119

$P_{CATG} = X/n = 0.9428$

Z = 1.96 (with a confidence level of 95%)

Applying the formula: $P_{CATG} \pm Z * [(P_{CATG} *(1-P_{CATG})/n)]^{0.5}$

We have:

= $0.9428 \pm (1.96) * (0.9428 * 0.0572 / 5429)^{0.5}$

= $0.9428 \pm (1.96) * (3.15 \times 10^{-3})$

= $0.9428 \pm 6.17 \times 10^{-3}$

Then, the confidence interval is as follows: **0.93663 <= p <= 0.94897**. This confidence interval is described in Figure 7:

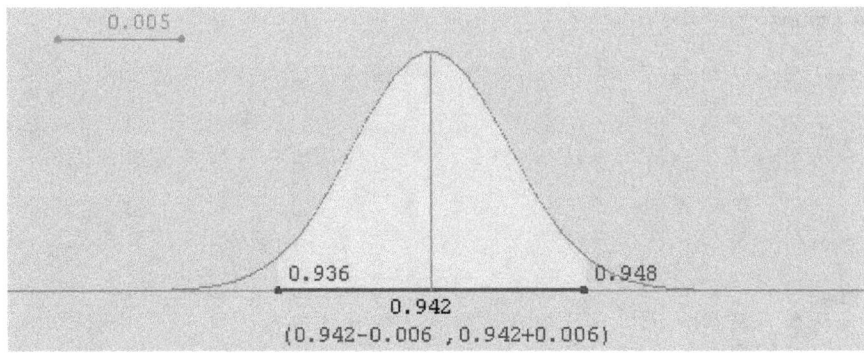

FIGURE 7. Confidence interval - Group CATG

Confidence interval calculation NCATG group

The data we have are:

n = 5429, X = 3555, $P_{NCATG} = X/n = 0.6548$, Z = 1.96 with a confidence level of 95%

Applying the formula:

$P_{NCATG} \pm Z * [(P_{NCATG} *(1-P_{NCATG})/n)]^{0.5}$

We have:

= 0.6548 ± (1.96) * (0.6548 * 0.3452 / 5429) ^ 0.5
= 0.6548 ± (1.96) * (6.45 x 10^{-3})
= 0.6548 ± 0.0126

The confidence interval is: **0.6422 <= p <= 0.6674,** which is described in Figure 8.

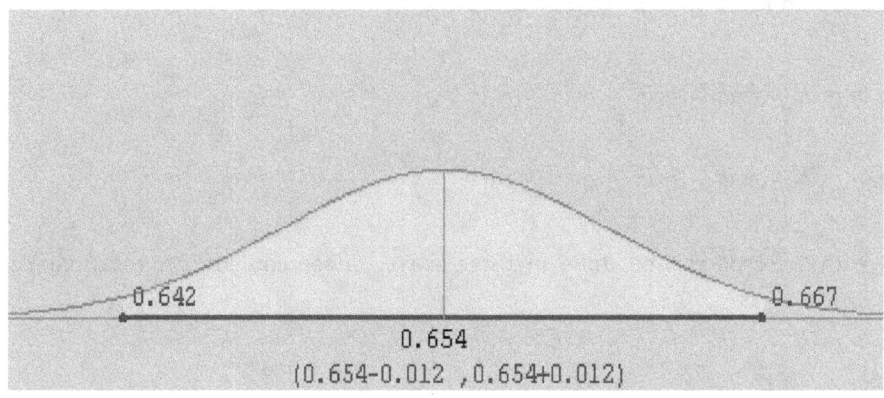

FIGURE 8. Confidence interval - Group NCATG

Hypothesis Testing

A situation that may be true or false on one or more people is called a *statistical hypothesis*. With the information extracted from the samples can test the hypotheses, but keep in mind that if accepted or rejected can make a mistake. The hypothesis formulated with the intention of rejecting it is called *null hypothesis* (H_0). Reject H_0 implies accepting the *alternative hypothesis* (H_1). The situation can be outlined:

TABLE 18. ERRORS HYPOTHESES

	H_0 true	H_0 false H_1 true
H_0 Reject	Type I Error (α)	Right decision (*)
H_0 no reject	Right decision	Type II Error (β)

Where:
α = probability (reject H_0, but, H_0 is true)
β = probability (acept H_0, but, H_0 is false)
(*) Correct decision is sought
$1-\beta$ = probability (reject H_0 where H_0 is false)

Details to consider

1. α and β are inversely related.
2. Only two can be reduced by increasing n.

The steps required to make a contrast on a parameter θ are:

1. Set the null hypothesis in terms of equality

$$H_0 : \theta = \theta_0$$

2. Set the alternative hypothesis that can be done in three ways, depending on the researcher's interest

$$H_1 : \theta \neq \theta_0 \qquad \theta > \theta_0 \qquad \theta < \theta_0$$

In the first case we speak of two-sided or two-tailed, and the other two lateral (right in the story 2 or the 3 rd left) or a tail.

3. Choose a significance level: critical level for α

4. Choose a statistic: statistic whose sampling distribution is known in H_0 and is related to θ and to establish on the basis of this distribution, the critical region: a region in which the statistic has a probability of less than α if H_0 is true and therefore, if the statistic falls in the same, would reject H_0.

Note that in this way is safer when a hypothesis is rejected or not. Therefore H0 is set as what they want to reject. When not rejected, it has not shown anything, just could not be rejected. On the other hand, the decision is based on the sampling distribution in H_0, so you must have equality.

5. Calculate the statistic for a random sample and compare it with the critical region, or equivalently, calculating the "p value" of the statistic (probability of obtaining that value, or other further from the H_0 if H_0 were true) and compare it to α.

Statistic to be used

The statistical test used is the difference between proportions for two independent populations using the Normal approximation [Berenson, 1996, Section 13.2. Formula 13.1. p. 439]:

$$Z = \frac{(p_{s1} - p_{s2}) - (p_1 - p_2)}{(P * (1-P) * (1/n1 + 1/n2))^{0.5}}$$

P = (X1 + X2) / (n1+n2) ; p_{s1} = X1 / n1 ; p_{s2} = X2 / n2

Where:

p_{s1} = Proportion of population 1
p_{s2} = Proportion of population 2
X1 = Hits of the population 1
X2 = Hits of the population 2
n1 = size of sample 1
n2 = size of sample 2
P = estimated combined proportion

Test of Hypothesis H1

Statement. The wording of the first hypothesis is as follows:

> "*The proportional difference in the development of skills among students using the CAT and not use it, when studying the Operating Systems Course in Computer Sciences Engineering program of the Faculty of Engineering at San Martin University Foundation headquarters Caribbean is 30%.*"

Construction of the null and alternative hypotheses.

The null and alternative hypotheses are:

$$Ho: P_{CATG} - P_{NCATG} = 0.3$$
$$H1: P_{CATG} - P_{NCATG} \neq 0.3$$

Selecting the significant level of α. *The significant level is $\alpha = 0.05$, i.e. you want a confidence level of 95%. We have thus the Z value of 1.96.*

Calculation of the rejection region.

With $\alpha = 0.05$ and Z = 1.96 the region to reject the null hypothesis of double tail is two areas:

$$Z > 1.96 \text{ o } Z < -1.96$$

30 Analysis of skills development from computer-assisted teaching

Performing the test of Hypotheses. It is necessary to replace the corresponding values in the selected statistic used, which found that:

$$Z = (0.9428 - 0.6548 - 0.3) / (0.7988 * 0.2011 * 0.000368)^{0.5}$$
$$Z = -0.01191748 / 0.007693807$$
$$Z = -1.548970595$$

We note that this value of Z (**1.548970595**)) is not in the area of rejection, may therefore <u>not reject</u> the null hypothesis (Ho : $P_{CATG} - P_{NCATG} = 0.3$). The above is described graphically in Figure 9.

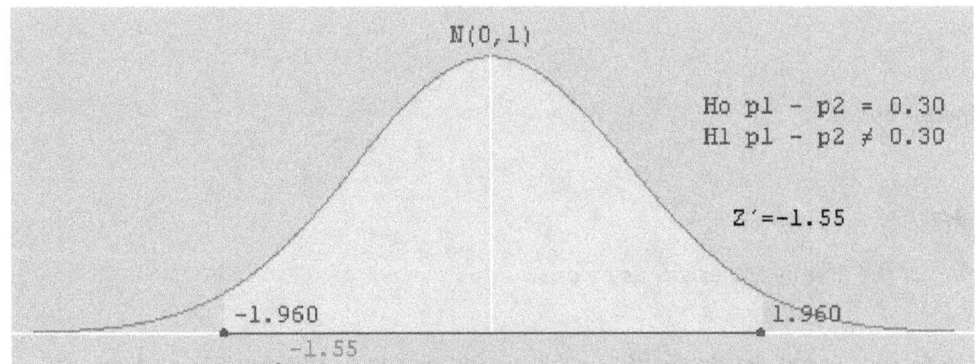

FIGURE 9. Hypothesis Testing $P_{CATG} - P_{NCATG} = 0.3$

Analysis of test results. In the previous section is concluded that cannot reject the hypothesis $P_{CATG} - P_{NCATG} = 0.3$. Then tested for a queue to see if the difference in proportions is more or less equal (see Table 19)

TABLE 19. ANALYSIS WITH P = 0.3

Hypothesis Ho	Hypothesis H1	p	Z	Rejection interval	Rejection
P1 - p2 = p	p1 - p2 ≠ p	0.3	-1.5489706	Z>1.96 o Z<-1.96	No
P1 - p2 >= p	p1 - p2 < p	0.3	-1.5489706	Z<-1.96	No
P1 - p2 <= p	p1 - p2 > p	0.3	-1.5489706	Z>1.96	No

It is not possible to reject any hypothesis (Ho), so more tests are needed at other intervals. The tests described in the above table are analyzed graphically in Figures 10 and 11.

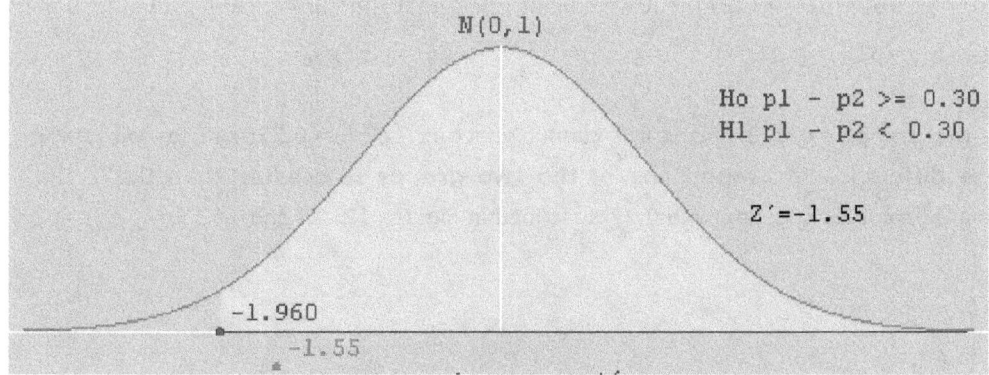

FIGURE 10. Hypothesis Testing P$_{CATG}$ - P$_{NCATG}$ >= 0.3

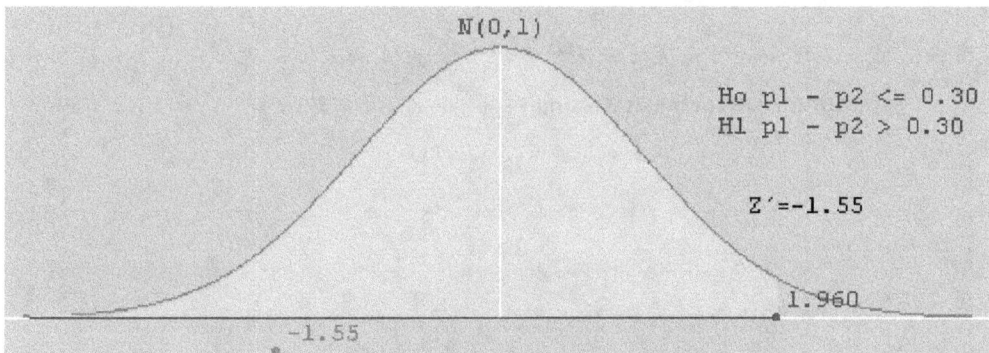

FIGURE 11. Hypothesis Testing P$_{CATG}$ - P$_{NCATG}$ <= 0.3

Now it is important to know the behavior around P = 0.3. Thus the first step will take as the difference in the proportion to the value 0.27, and we apply the hypothesis testing. In doing so, we get two of three rejections as acacia in Table 24.

TABLE 20. ANALYSIS WITH P=0.27

Hypothesis Ho	Hypothesis H1	p	Z	Rejection interval	Rejection
P1 - p2 = p	p1 - p2 ≠ p	0.27	2.3502696	Z>1.96 o Z<-1.96	Yes
P1 - p2 >= p	p1 - p2 < p	0.27	2.3502696	Z<-1.96	**No**
P1 - p2 <= p	p1 - p2 > p	0.27	2.3502696	Z>1.96	Yes

Based on the information provided in the table above, we can accept that the difference in proportions, is not equal to 0.27nor less; because these hypotheses were rejected (and accepted the alternative

H1: p1 - p2 ≠ 0.27 y H1: p1 - p2 > 0.27), but we cannot reject that the difference in proportions of the two populations is p1 - p2> = 0.27.

Now, if it is accept that H1: p1 - p2> 0.27 and it is cannot reject p1 - p2> = 0.27, can say with 95% confidence that **the difference in proportions of the two groups is greater than 0.27.** The graphic description of a hypothesis testing with 0.27 described in Figures 12, 13 and 14.

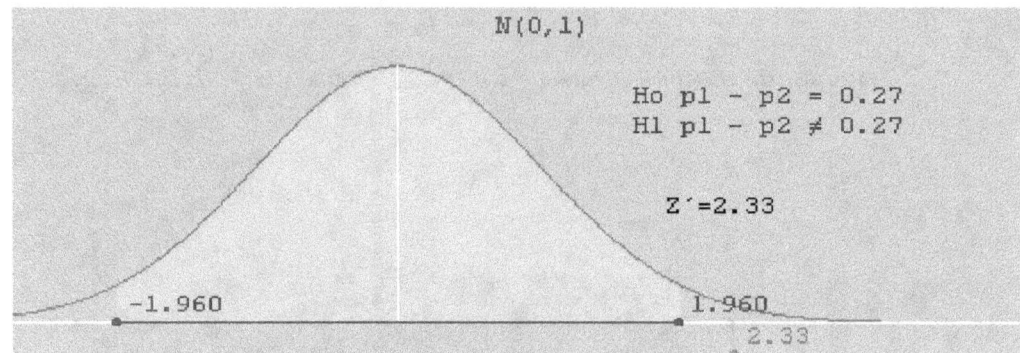

FIGURE 12. Hypothesis Testing $P_{CATG} - P_{NCATG} = 0.27$

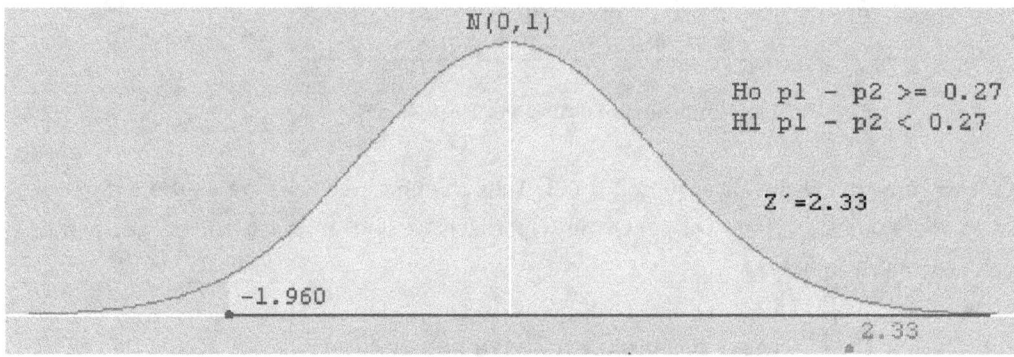

FIGURE 13. Hypothesis Testing $P_{CATG} - P_{NCATG} >= 0.27$

Similarly, will make hypothesis testing with a value greater than 0.3, and take a very close to it as it is 0.31. By doing all three (3) presents two tests rejects the null hypothesis. The first indication that the difference cannot be equal to 0.31, thereby accepting the alternative hypothesis that expresses the inequality. And the second, which indicates the impossibility of that difference is greater than or equal, **it is accepted that the difference is less than 0.31**. Table 21 shows the test results

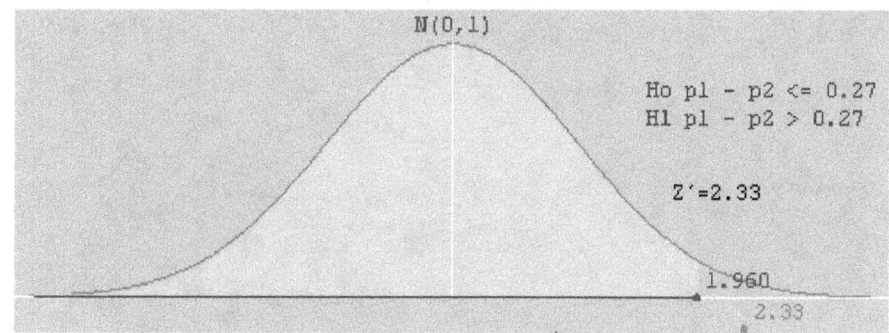

FIGURE 14. Hypothesis Testing $P_{CATG} - P_{NCATG} <= 0.27$

TABLE 21. ANALYSIS WITH P=0.31

Hypothesis Ho	Hypothesis H1	p	Z	Rejection interval	Rejection
P1 - p2 = p	p1 - p2 ≠ p	0.31	-2.84871733	Z>1.96 o Z<-1.96	Yes
P1 - p2 >= p	p1 - p2 < p	0.31	-2.84871733	Z<-1.96	Yes
P1 - p2 <= p	p1 - p2 > p	0.31	-2.84871733	Z>1.96	**No**

It is clear that by accepting the alternative hypothesis states, in part, the failure equal to 0.31, and in part the difference is less at 0.31, and unable to reject the null hypothesis Ho: P1-p2 <= 0.31, can be conclude that **the value of the difference in proportions of the groups is less than 0.31**. Graphic description of the hypothesis tests of 0.31 is made in Figures 15, 16 and 17.

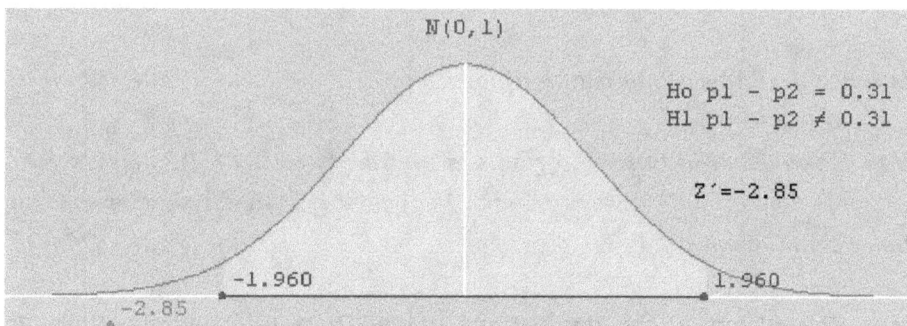

FIGURE 15. Hypothesis Testing $P_{CATG} - P_{NCATG} = 0.31$

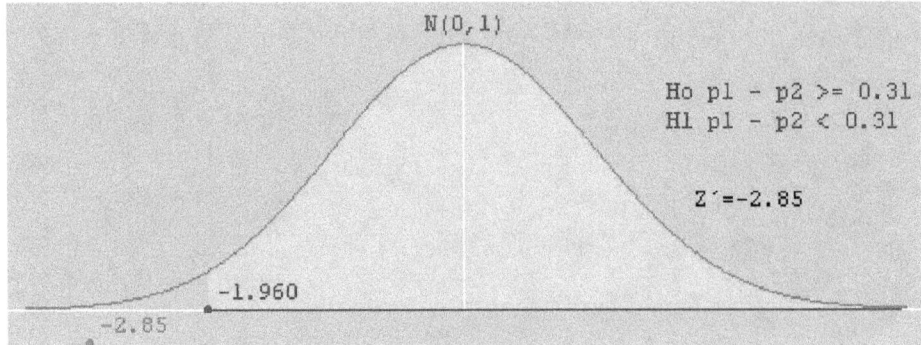

FIGURE 16. Hypothesis Testing P$_{CATG}$ - P$_{NCATG}$ >= 0.31

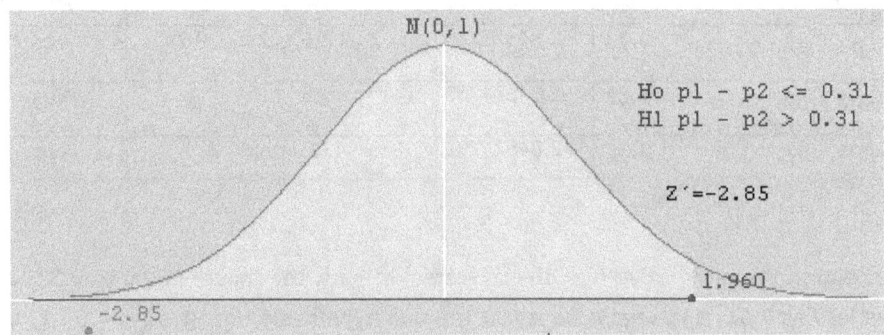

FIGURE 17. Hypothesis Testing P$_{CATG}$ - P$_{NCATG}$ <= 0.31

In summary, we have accepted the following Alternative hypotheses:

a) **The difference in proportions of the two groups is greater than 0.27.** It is argued that values less than or equal to 0.27 always accept the hypothesis that the difference in proportions of the groups is greater

b) **The difference in proportions of the two groups is less than 0.31.** Indicates that for values greater than 0.31 provided the difference in proportions will be less.

c) And on the other side is that cannot reject the null hypotheses that are described in Table 22

TABLE 22. NO REJECTION ANALYSIS WITH DIFFERENT VALUES OF P

Hypothesis Ho	Hypothesis H1	p	Z	Rejection interval	Rejection
P1 - p2 = p	p1 - p2 ≠ p	0.300	-1.5489706	Z>1.96 o Z<-1.96	No
P1 - p2 >= p	p1 - p2 < p	0.300	-1.5489706	Z<-1.96	No
P1 - p2 <= p	p1 - p2 > p	0.300	-1.5489706	Z>1.96	No
P1 - p2 <= p	p1 - p2 > p	0.310	-2.84871733	Z>1.96	No
P1 - p2 >= p	p1 - p2 < p	0.270	2.3502696	Z<-1.96	No
P1 - p2 <= p	p1 - p2 > p	0.305	-2.19884396	Z>1.96	No
P1 - p2 = p	p1 - p2 ≠ p	0.280	1.05052287	Z>1.96 o Z<-1.96	No
P1 - p2 >= p	p1 - p2 < p	0.280	1.05052287	Z<-1.96	No
P1 - p2 <= p	p1 - p2 > p	0.280	1.05052287	Z>1.96	No
P1 - p2 = p	p1 - p2 ≠ p	0.301	-1.67894527	Z>1.96 o Z<-1.96	No
P1 - p2 >= p	p1 - p2 < p	0.301	-1.67894527	Z<-1.96	No
P1 - p2 <= p	p1 - p2 > p	0.301	-1.67894527	Z>1.96	No
P1 - p2 <= p	p1 - p2 > p	0.400	-14.5464379	Z>1.96	No
P1 - p2 >= p	p1 - p2 < p	0.100	24.4459641	Z<-1.96	No

Following from the above analysis, this research project is accepted with a 95% confidence that:

The difference in proportions between groups CATG, and NCATG is 0.3, which is equivalent to 30%.

ANALYSIS OF INFORMATION PROVIDED BY THE ACADEMIC SECRETARY

Data description and summary

The data obtained by carrying out partial notes in each group was calculated the mean, its variance and standard deviation, which are summarized in Table 23.

TABLE 23. SUMMARY DATA

Group	Mean	Variance	Deviation
CATG	3.660	0.292	0.541
NCATG	3.185	0.449	0.670

Other measures of central tendency and dispersion are listed in Table 24:

TABLE 24. MEASURES OF CENTRAL AND DISPERSION

Group	Median	Mode	Half range	Range
CATG	3.580	3.050, 3.220, 3.320	3.47	2.00 - 4.94
NCATG	3.100	3.040, 3.020	2.35	0.00 – 4.69

While the frequency distribution CATG group is shown in Figure 18, the frequency distribution NCATG Group is presented in Figure 19.

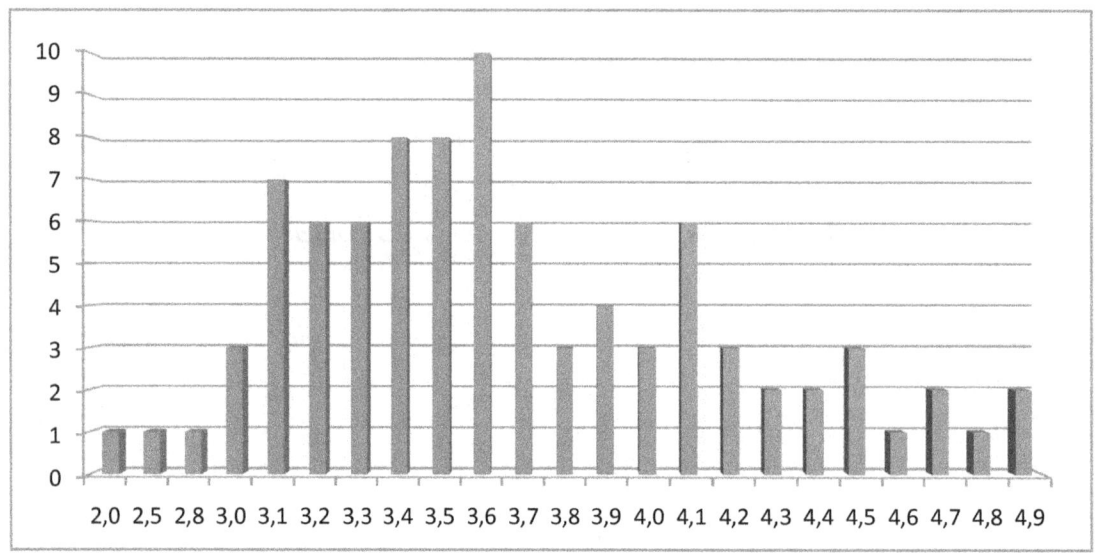

FIGURE 18. CATG Group Frequency Distribution

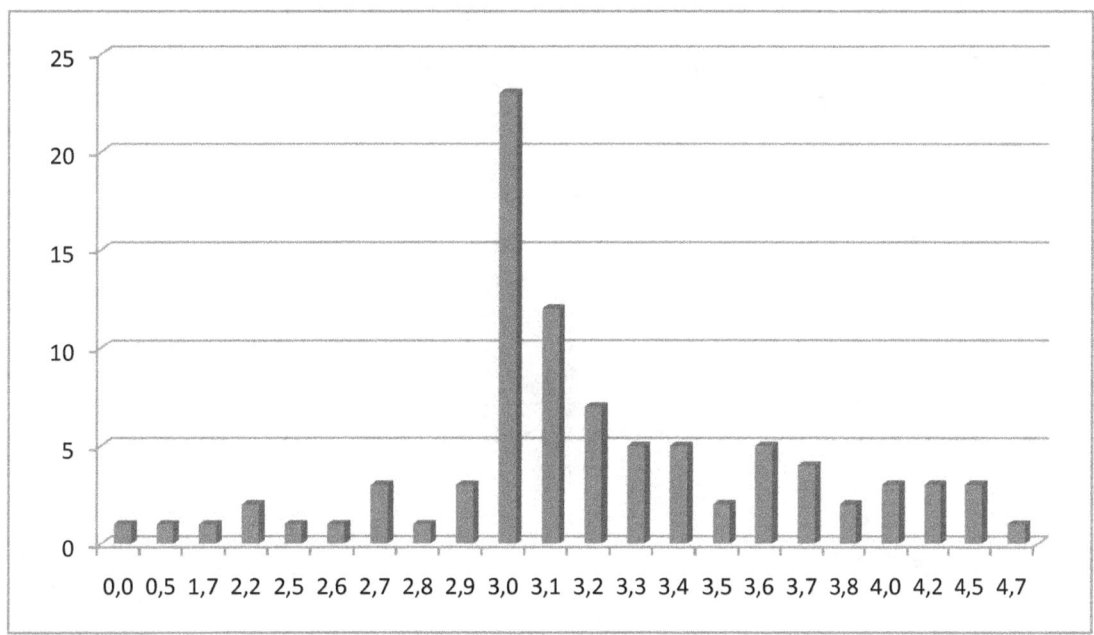

FIGURE 19. NCATG Group Frequency Distribution

Calculation of Confidence Intervals

Used statistical

To calculate the confidence intervals of the proportions of the groups CATG, and NCATG use the formula for confidence interval estimation of the mean with unknown population mean [Berenson, 1996, Section 10.3. p. 334] based on the Student T distribution.

Probability t, or T Student. A random variable is distributed according to the probability model t, or T Student with k degrees of freedom, where k is a positive integer, if its density function is as follows [López Sánchez, 2010]:

$$f(t) = \frac{\Gamma(\frac{k+1}{2})}{\sqrt{\pi k}\,\Gamma(\frac{k}{2})} (1+\frac{t^2}{k})^{-(\frac{k+1}{2})}, \ -\infty < t < \infty, \ \text{donde } \Gamma(p) = \int_0^\infty e^{-x} x^{p-1} dx$$

The graph of this density function is symmetrical about the vertical axis, irrespective of the value of k, and somewhat similar to that of a normal distribution. (See Figure 20)

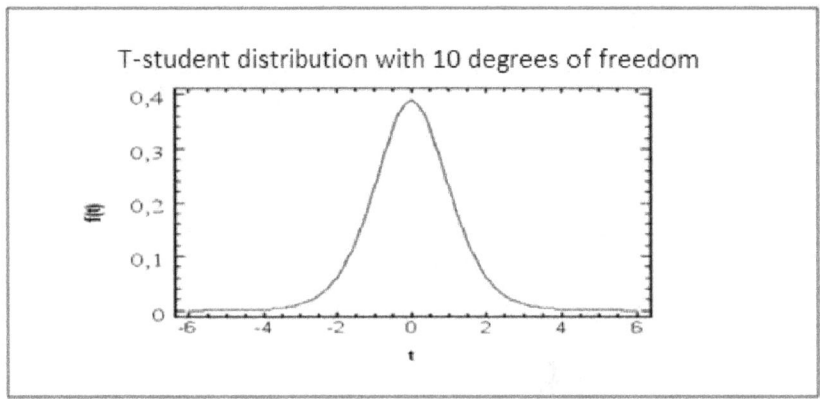

FIGURE 20. Distribution T STUDENT [López Sánchez, 2010]

Its mean and variance are [López Sánchez, 2010]:

$$E(T)=\mu=\int_{-\infty}^{\infty} tf(t)dt=\int_{-\infty}^{\infty} t\frac{\Gamma(\frac{k+1}{2})}{\sqrt{\pi k}\Gamma(\frac{k}{2})}(1+\frac{t^2}{k})^{-(\frac{k+1}{2})}dt=\ldots=0$$

Si $k>2$,

$$Var(T)=\sigma^2=E((T-\mu)^2)=\int_{-\infty}^{\infty}(t-\mu)^2\frac{\Gamma(\frac{k+1}{2})}{\sqrt{\pi k}\Gamma(\frac{k}{2})}(1+\frac{t^2}{k})^{-(\frac{k+1}{2})}dt=\ldots=\frac{k}{k-2}$$

The law of probability of the sample mean in a normal population with unknown variance. If X1, X2, ..., Xn are independent random variables with normal probability law N (μ, σ), i.e., a random sample of size n drawn from a population N (μ, σ), then [López Sánchez, 2010]

$\dfrac{\bar{X}-\mu}{\frac{S}{\sqrt{n}}}$ Is distributed as a t-student variable

with (n-1) degrees of freedom, where $S^2 = \sum_{i=1}^{n}\dfrac{(X_i-\bar{X})^2}{n-1}$

is the sample variance and $\bar{X}=\sum_{i=1}^{n}\dfrac{X_i}{n}$ is the sample mean.

The statistic to be used is as follows:

Sample Mean $\pm$ t$_{n-1}$ (standard deviation / (sample size) ^ 0.5)

Confidence interval calculation of CATG group

It is found that:

n = 89; X_{CATG} = 3.66 ; T_{88} = 1.9873 (with a confidence level of 95% and 88 degrees of freedom); S_{CATG} = 0.541

Applying the formula: $X_{CATG} \pm T_{88} * S_{CATG} / (n)^{0.5}$

= 3.66 ± (1.9873) * 0.541/ (89) ^ 0.5
= 3.66 ± 1.075 / 9.43
= 3.66 ± 0.1139

Then, the confidence interval is: **3.5461 <= U_{CATG} <= 3.7739**. This confidence interval is described in the following Figure 21:

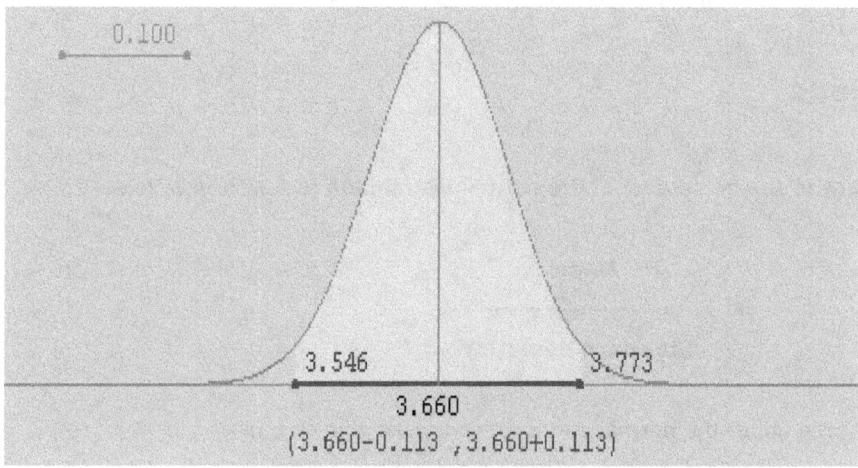

FIGURE 21. Confidence interval of CATG Group

Confidence interval calculation of NCATG group

In this case it has to: n = 89; X_{NCATG} = 3.185; T_{88} = 1.9873 (with a confidence level of 95% and 88 degrees of freedom); S_{NCATG} = 0.670

Applying the formula: $X_{NCATG} \pm T_{88} * S_{NCATG} / (n)^{0.5}$

= 3.185 ± (1.9873) * 0.670 / (89) ^ 0.5
= 3.185 ± 1.331 / 9.43
= 3.185 ± 0.141

Then, the confidence interval is: **3.044 <= U_{NCATG} <= 3.326** This confidence interval is described in the following Figure 22:

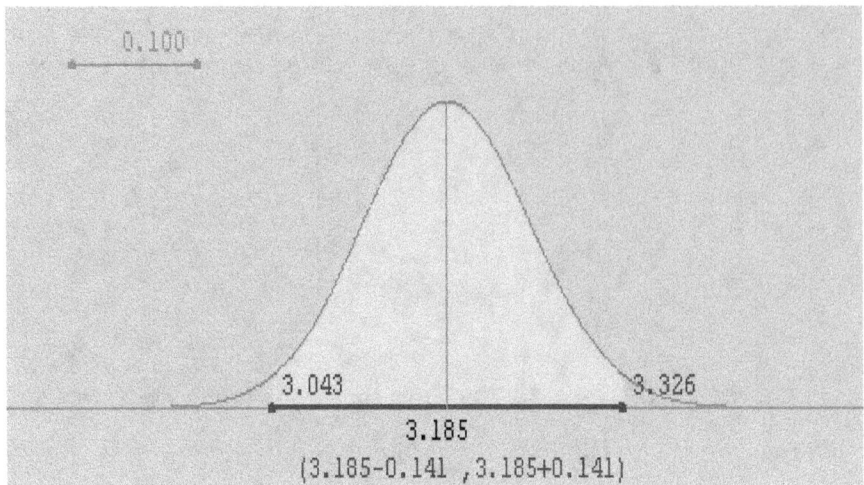
FIGURE 22. Confidence interval of NCATG Group

Comparison populations

Use statistical

Formula will be used to standardize the processing of the normal distribution [Berenson, 1996, Section 8.2.4. Page 269] whose formula is:

$$Z = \frac{X - \text{Mean}}{\text{Standard deviation}}$$

Then the value of this Z is the area under the normal curve with the following formula:

$$f(z;0,1) = \frac{1}{\sqrt{2\pi}} e^{-\frac{z^2}{2}}$$

Good performance compare

A good performance in a course anyone to assume that the student received a grade higher or equal to 4.0. Therefore, for each group perform the test and then compare the area above the normal curve.

For the group CATG: It is found that has a mean = 3.66, standard deviation = 0.541, then:
  Z = (4.0 − 3.66) / 0.541 = 0.34 / 0.541
  Z = 0.62846

And the value F (Z) = F (0.62846) = **0.2351**. To find the upper area would subtract 0.5 and result **0.2649**. Figure 23 shows the process graphically:

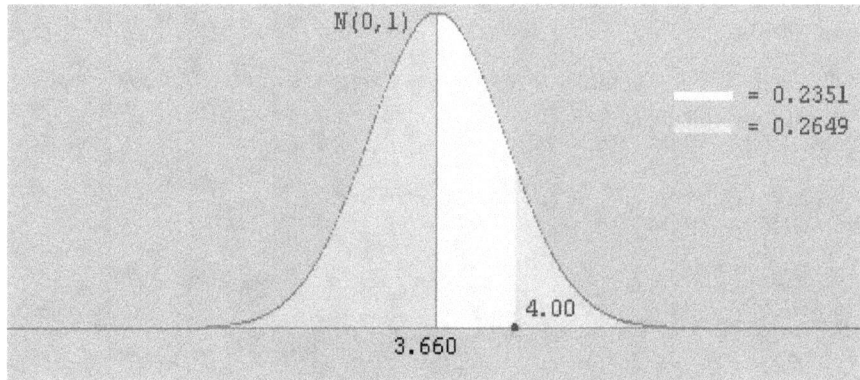

FIGURE 23. Good Area for CATG Group

For the group CATG: It is found that has a mean = 3.185, standard deviation = 0.670, then:

Z = (4.0 − 3.185) / 0.670 = 0.815 / 0.670

Z = 1.2164

And the value F (Z) = F(1.21641) = **0.3880**, to find the upper area would subtract 0.5 and result **0.1119**. Figure 24 shows the process graphically:

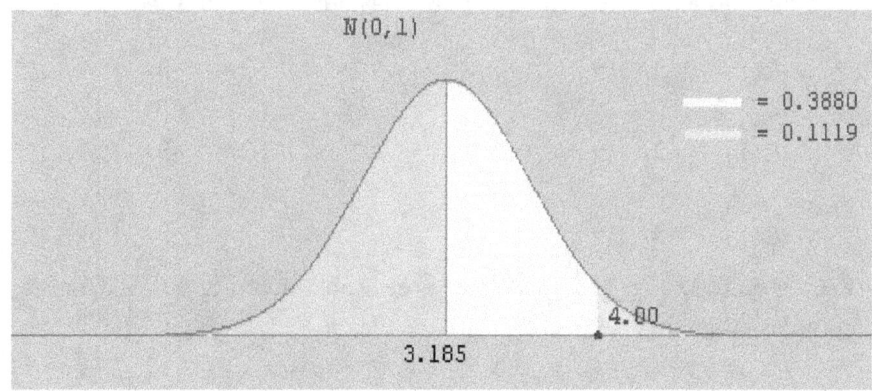

FIGURE 24. Good area for NCATG Group

Poor performance compare

A poor performance in a course anyone to assume that the student received a grade below 3.0. Therefore, for each group perform the test and then compare the area above the normal curve.

For the group CATG: It is found that has a mean = 3.66, standard deviation = 0.541, then:

Z = (3.0 − 3.66) / 0.541
Z = −0.66 / 0.541
Z = −1.2199

And the value F(Z) = F(−1.2199) = **0.3887**. To find the upper area would subtract 0.5 and result **0.1113**. Figure 25 shows the process graphically.

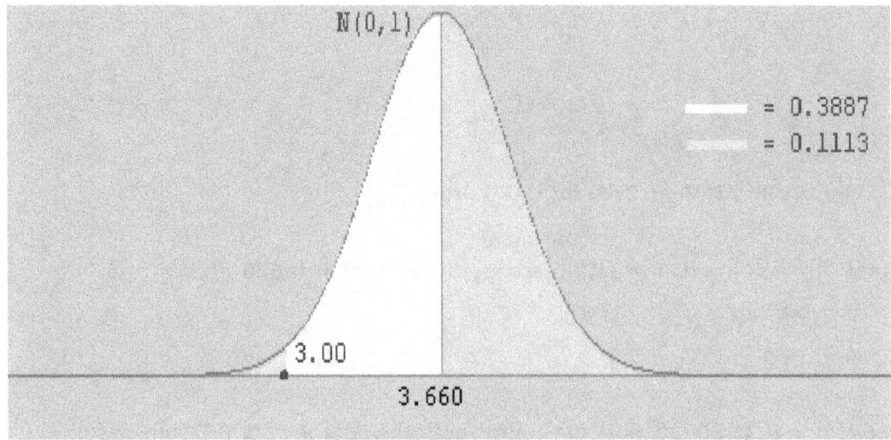

FIGURE 25. Poor Area for CATG Group

For the group CATG: It is found that has a mean = 3.185, standard deviation = 0.670, then:

Z = (3.0 − 3.185) / 0.670
Z = −0.1815 / 0.670
Z = −0.276

And the value F(Z) = F(−0.276) = **0.1087**. To find the upper area would subtract 0.5 and result **0.3913**. Figure 26 shows the process graphically:

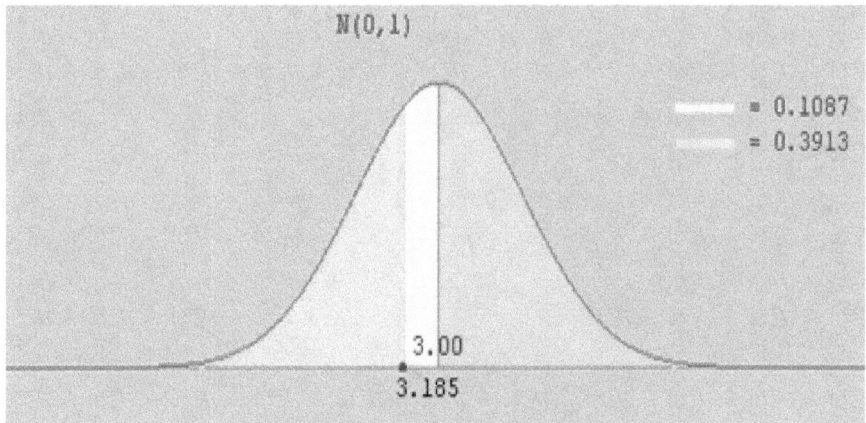

FIGURE 26. Poor Area for NCATG Group

Finally, Figure 27 is described in the form of bars compare differences between good and poor performance of the groups:

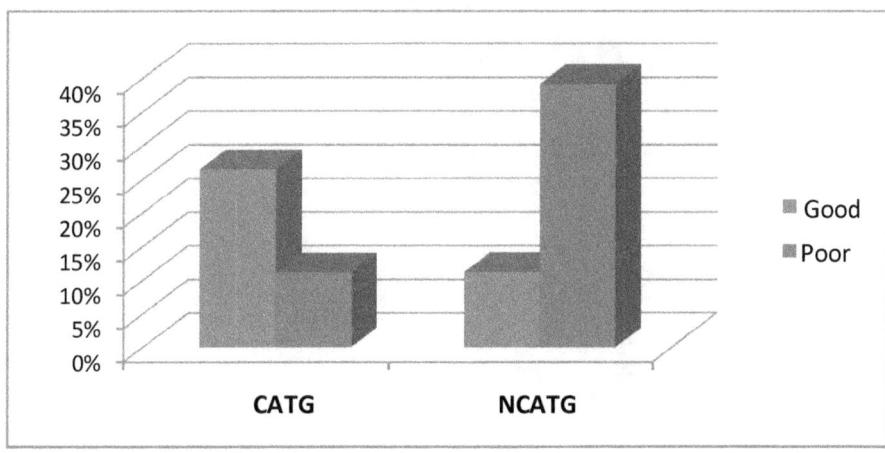

FIGURE 27. Performance comparison CATG vs. NCATG

CONCLUSIONS AND RECOMMENDATIONS

From hypothesis testing, we can say, first, that the development of skills in the field of computer science Engineering, is 30% higher when using computer-assisted teaching and second That the level of students who get good academic performance is higher with the use of CAT methodology.

At present the traditional way in the engineering faculties of universities, high mortality rates in subjects in the area of basic sciences (mathematics, physics, etc..), it is appropriate, taking the results, encourage the development and use of educational software in this area.

Finally, computer-assisted teaching, attempting to develop practical skills, brings us a bit of that reality that needs professional, and the person is competent in the world today. Also, if you contribute to the development of skills in a 30% increase, would avoid the new "professional incompetence" and thus human welfare also increases.

From the results it can be stated the following recommendations:

- Using the methodology of computer-assisted teaching in areas such as mathematics, and specific training in Systems Engineering program.
- Development of educational software projects that nurture the subjects of the different programs offered by the foundation.
- Integrate computer-assisted teaching on the culture of the institution.
- Contextualizing the courses in the academic programs based on the CAT.

BIBLIOGRAPHY

[Aedo et al., 2004]   Aedo, I, Díaz, P., Sicilia, M.A., Colmenar, A., Losada, P., Mur, F., Castro, M. y Peire, J. (2004): Sistemas multimedia: análisis, diseño y evaluación. Editorial UNED. En Díaz, M, Montero, S & Aedo, I. (2005) Ingeniería Web y patrones de diseño. Universidad Carlos III Madrid. Prentice – Hall, Madrid. P 11.

[Aproa, 2007]   APROA Comunidad (2007) ¿Qué es un Objeto de Aprendizaje? Proyecto FONDEF. Aprendiendo con Repositorio de Objetos de Aprendizaje.Centro Agrimed, Universidad de Chile [On-Line], Aviliable: http://www.aproa.cl/1116/propertyvalue-5538.html

[Berenson, 1996]   Berenson, Mark and Levine, David. (1996) Estadística básica en administración: Conceptos y aplicaciones.4 Ed. Prentice – Hall, México. 946 p.

[Bertoa, Troya, & Vallecillo, 2002]   Bertoa, Manuel F., Troya, José M. y Vallecillo, Antonio. (2002). Aspectos de Calidad en el Desarrollo de Software Basado en Componentes. Depto. Lenguajes y Ciencias de la Computación. Universidad de Málaga. [On-Line], Aviliable: http://www.lcc.uma.es/~av/Publicaciones/02/CalidadDSBC.pdf

[Casal, J., 2007]   Casal, J. (2007) Microsoft Desarrollo de Software basado en Componentes. [On-Line], Aviliable: http://www.microsoft.com/spanish/msdn/comunidad/mtj.net/voices/

[Cataldi, Z., et al., 2002]   Cataldi, Zulma et al. (2002) Metodología extendida para la creación de software educativo desde una visión integradora. Revista latinoamericana de tecnología educativa volumen 2 número 1.

[Ceri, Fraternali, and Bongio, 2000]   Ceri, Stefano, Fraternali, Piero and Bongio, Aldo (2000).Web Modeling Language (WebML): a modeling language for designing Web sites. [On-Line], Aviliable: www.webml.org/webml/upload/ent5/1/www9.pdf

[Díaz de Feijoo, M., 2002]   Díaz de Feijoo., María Gabriela (2002). Propuesta de una metodología de desarrollo y evaluación de software educativo bajo un enfoque de calidad sistémica. Tesis de Especialización. Universidad Simón Bolívar.

[Díaz, Aedo, y Montero., 2001]   Díaz, P., Aedo, I. y Montero, S. (2001). Ariadne, a development method for hypermedia. In proceedings of Dexa 2001, volume 2113 of Lecture Notes in Computer Science, pages 764-774,

[Díaz, Montero, & Aedo, 2005]   Díaz, M, Montero, S & Aedo, I. (2005) Ingeniería Web y patrones de diseño. Universidad Carlos III Madrid. Prentice – Hall, Madrid. 409 p.

[DoD.,1987]   DoD (1987). Report of the defense Science Board Task

	Force on Military Software. Departamento de Defensa de los Estados Unidos 1987 [On-Line], Aviliable:http://www.acq.osd.mil/dsb/reports/defensesoftware.pdf
[Douglass, B. , 1999]	Douglass, B. (1999) Doing Hard Time; Developing Real-Time Systems with UML, Objects, Frameworks, and Patterns. Addison-Wesley, United States of America. 749 p.
[Eyssautier, M., 2002]	Eyssautier De La Mora, Maurice (2002). Metodología de la Investigación: Desarrollo de la Inteligencia. 4 Ed. Thompsom Editores. México. 316 p.
[Fernández ,Luís., 2000]	Fernández Sanz, Luís (2000). El futuro de la ingeniería del software o ¿cuándo será el software un producto de ingeniería?, Novática, nº 145, mayo-Junio, 2000, p. 82 77 [On-Line], Aviliable: http://www.ati.es/novatica/2000/145/luifer-145.pdf
[Friesen, N.,2001]	Friesen, N. (2001). What are educational objects?. Interactive learning environments, Vol. 9, No. 3, pp. 219-230.
[Friss de Kereki, I., 2003]	Friss de Kereki Guerrero., Inés (2003). Modelo para la Creación de Entornos de Aprendizaje basados en técnicas de Gestión del Conocimiento. Tesis Doctoral. Universidad Politécnica de Madrid. Madrid, España.
[García E. et al., 2002]	Garcia Roselló, E. et al. (2002)¿Existe una situación de crisis del software educativo?. VI Congreso Iberoamericano de Informática Educativa. [On-Line], Aviliable:http://lsm.dei.uc.pt/ribie/docfiles/txt2003729185619paper-144.pdf
[Gómez, Galvis y Mariño, 1998]	Gómez Castro, R., Galvis Panqueva, A. y Mariño Drews, O. Ingeniería del software educativo con modelaje orientado por objetos : un medio para desarrollar micromundos interactivos. Informática Educativa, Uniandes – Lidie, Vol 11, No 1,1998, pp.9-30.
[Gould, Boies y Ukelson. , 1997]	J. D. Gould, S. J. Boies y J. P. Ukelson. (1997) «How to design usable systems». En Handbook of Human Computer Interaction, pp 231-254. Elsevier Science, 1997. En Díaz, M, Montero, S & Aedo, I. (2005) Ingeniería Web y patrones de diseño. Universidad Carlos III Madrid. Prentice – Hall, Madrid. P 16.
[Hermans and De Vries,2006]	Hermans, H. and De Vries, F. (2006) Organizational scenario's for the use of learning objects. Learning Objects in practice 2. Stichting Digitale Universiteit. Netherlands

[Hurtado, Dougglas., 2007]	Hurtado Carmona, Dougglas, (2007). Análisis del desarrollo de competencias desde la enseñanza asistida por computador In: VI Encuentro iberoamericano de instituciones de enseñanza de la ingeniería XXVII Reunión ACOFI, 2007, Cartagena: El profesor de Ingeniería. Profesional de la formación de Ingenieros. p.112. ISSN 978-958-68005-5-6
[Hurtado y Neira, 1995]	Hurtado, Dougglas y Neira, Marlon. Software aplicativo para la enseñanza de la asignatura Sistemas Operacionales. Tesis de Grado. Universidad del Norte, Barranquilla, 1995. 248 p.
[Iglesias, C., 1998]	Iglesias, C. (1998).Definición de una metodología para el desarrollo de sistemas multiagentes. Tesis Doctoral, Universidad Politécnica de Madrid, España. 294 p.
[Kendall and Kendall., 1997]	Kendall, kenneth. Kendall, julie. (1997) Análisis y diseño de sistema. Pentice-hall. 913 p
[López Sánchez, 2010]	López Sánchez, Jesús et al. Probabilidad t o T de Student. Universidad Complutense de Madrid, españa. Dpto. de Matemática Aplicada. Proyecto de Innovación Educativa. Disponible en: http://e-stadistica.bio.ucm.es/glosario2/distr_student.html
[Mendoza, P., Galvis , A., 1999]	Mendoza B., Patricia. Galvis P., Alvaro.(1999) Ambientes virtuales de aprendizaje: una metodología para su creación. Revista Informática Educativa Vol 12, No, 2, 1999. Uniandes - Lidie pp.295-317
[Milenkovic, 1997]	Milenkovic, Milan. Sistemas operativos, conceptos y diseño". Mc Graw Hill Hispanoamericana de España, 1997.
[MinEdu y Ciencia-GobEspañol, 2007]	Ministerio de Educación y Ciencia, Gobierno Español. (2007). Descartes. Software de apoyo gráfico estadístico. Consultado el 21 de mayo de 2007 en http://descartes.cnice.mec.es/Descartes1/index.html; http://recursostic.educacion.es/descartes/web/
[Montero, Díaz & Aedo, 2006]	Montero, Díaz, M, S & Aedo, I. (2006) ADM: Método de diseño para la generación de prototipos web rápidos a partir de modelos. XV Jornadas de Ingeniería del Software y Bases de Datos JISBD 2006 José Riquelme - Pere Botella (Eds) Ó CIMNE, Barcelona, 2006. [On-Line], Aviliable:http://www.dsic.upv.es/workshops/dsdm06/files/dsdm06-03-Montero.pdf
[Naranjo, 2005]	Naranjo, Fernando.(2005). Calidad de software. Escuela Universitaria Politécnica de Teruel.
[Nieto-Santisteban, 2001]	Nieto-Santisteban, María A. (2001). Ingeniería Web. Construyendo Web Apps. I Jornadas de Ingeniería Web' 01. [On-Line], Aviliable: http://www.informandote.com/jornadasIngWEB/articulos/jiw01.pdf

[Novática, 1996]	Anónimo. Si los programadores fueran albañiles... Novática, nº 124, noviembre-diciembre, 1996, p. 77 [On-Line], Aviliable: http://www.ati.es/novatica/1996/124/if124.html
[Pressman., 2002]	Pressman, Roger. (2002). Ingeniería del software: un enfoque práctico. 5 ed. México: McGraw-Hill. Madrid. 601 p.
[Sametinger, J. , 1997]	Sametinger, J. (1997) Software Engineering with Reusable Components. Berlin: Springer.
[Sanz, Aedo y Díaz., 2006]	Sanz, Daniel, Aedo, Ignacio y Díaz, Paloma (2006). Un Servicio Web de Políticas de Acceso Basadas en Roles para Hipermedia. [On-Line], Aviliable: http://www.ewh.ieee.org/reg/9/etrans/vol4issue2April2006/4TLA2_3Sanz.pdf
[Shaw, 1994]	Shaw,M.,(1994). Prospects for an engineering discipline of software En: J. Marciniak (ed.), Software Engineering Encyclopedia, IEEE, 1994, pp. 930-940.
[Silberschatz, 2006]	Silberschatz. Abraham, Gagne Greg, Galvin Peter Baer. Fundamentos de sistemas operativos. 7a. ed. México, Mcgraw-hill, 2006.
[Stallings, 2005]	Stallings, William. Sistemas operativos: aspectos internos y principios de diseño . 5a. ed. Madrid, Pearson Educación, 2005.
[Stallings. 2001]	Stallings, William. Sistemas operativos: principios de diseño e interioridades. 4a. ed. Madrid, Pearson Educación, 2001.
[Tanenbaum, 2003]	Tanenbaum, Andrew S. Sistemas operativos modernos". 2a. ed. México, Pearson Educación, 2003.
[Vargas, M., 2007]	Vargas, María Leonor. Repositorios de Objetos de Aprendizaje. [On-Line],. Visitada 09/03/2007Aviliable:http://www.alejandria.cl/recursos/documentos/documento_varas.doc.
[Wiley, D., 2000]	Wiley, David.(2000). Learning Object Design and Sequencing Theory. Tesis doctoral no publicada de la Brigham Young University. Accesible en http://davidwiley.com/papers/dissertation/dissertation.pdf
[Wiley, D. 2001]	Wiley, D. (2001). Connecting learning objects to instructional design theory: A definition, a methaphor, and a taxonomy.
[Wiley, D. , 2006]	Wiley, D. (2006) R.I.P. ping on Learning Objects. [On-Line], Aviliable: http://opencontent.org/blog/archives/230